TECHNOLOGY
and SOCIETY

OPPOSING VIEWPOINTS®

Other Books of Related Interest

TECHNOLOGY
and SOCIETY

OPPOSING VIEWPOINTS®

Auriana Ojeda, *Book Editor*

Daniel Leone, *President*
Bonnie Szumski, *Publisher*
Scott Barbour, *Managing Editor*

**OPPOSING
VIEWPOINTS®
SERIES**

GREENHAVEN PRESS
SAN DIEGO, CALIFORNIA

GALE GROUP

THOMSON LEARNING
Detroit • New York • San Diego • San Francisco
Boston • New Haven, Conn. • Waterville, Maine
London • Munich

4/02

47971844

Cover photo: Digital Stock

Library of Congress Cataloging-in-Publication Data

Technology and society: opposing viewpoints / Auriana Ojeda,
book editor.
    p. cm. — (Opposing viewpoints series)
    Includes bibliographical references and index.
    ISBN 0-7377-0912-X (pbk. : alk. paper) —
ISBN 0-7377-0913-8 (lib. : alk. paper)
    1. Technology—Social aspects. 2. Technological
innovations—Social aspects. I. Ojeda, Auriana, 1977–
II. Opposing viewpoints series (Unnumbered)

HM846 .T43 2002
306.4'6—dc21   2001050185

"Congress shall make no law...abridging the freedom of speech, or of the press."

First Amendment to the U.S. Constitution

The basic foundation of our democracy is the First Amendment guarantee of freedom of expression. The Opposing Viewpoints Series is dedicated to the concept of this basic freedom and the idea that it is more important to practice it than to enshrine it.

Contents

Chapter 3: How Has Technology Affected Privacy?

Chapter 4: How Will Technology Affect Society in the Future?

Why Consider Opposing Viewpoints?

"The only way in which a human being can make some approach to knowing the whole of a subject is by hearing what can be said about it by persons of every variety of opinion and studying all modes in which it can be looked at by every character of mind. No wise man ever acquired his wisdom in any mode but this."

John Stuart Mill

In our media-intensive culture it is not difficult to find differing opinions. Thousands of newspapers and magazines and dozens of radio and television talk shows resound with differing points of view. The difficulty lies in deciding which opinion to agree with and which "experts" seem the most credible. The more inundated we become with differing opinions and claims, the more essential it is to hone critical reading and thinking skills to evaluate these ideas. Opposing Viewpoints books address this problem directly by presenting stimulating debates that can be used to enhance and teach these skills. The varied opinions contained in each book examine many different aspects of a single issue. While examining these conveniently edited opposing views, readers can develop critical thinking skills such as the ability to compare and contrast authors' credibility, facts, argumentation styles, use of persuasive techniques, and other stylistic tools. In short, the Opposing Viewpoints Series is an ideal way to attain the higher-level thinking and reading skills so essential in a culture of diverse and contradictory opinions.

In addition to providing a tool for critical thinking, Opposing Viewpoints books challenge readers to question their own strongly held opinions and assumptions. Most people form their opinions on the basis of upbringing, peer pressure, and personal, cultural, or professional bias. By reading carefully balanced opposing views, readers must directly confront new ideas as well as the opinions of those with whom they disagree. This is not to simplistically argue that everyone who reads opposing views will—or should—change his or her opinion. Instead, the series enhances readers' understanding of their own views by encouraging confrontation with opposing ideas. Careful examination of others' views can lead to the readers' understanding of the logical inconsistencies in their own opinions, perspective on why they hold an opinion, and the consideration of the possibility that their opinion requires further evaluation.

Evaluating Other Opinions

To ensure that this type of examination occurs, Opposing Viewpoints books present all types of opinions. Prominent spokespeople on different sides of each issue as well as well-known professionals from many disciplines challenge the reader. An additional goal of the series is to provide a forum for other, less known, or even unpopular viewpoints. The opinion of an ordinary person who has had to make the decision to cut off life support from a terminally ill relative, for example, may be just as valuable and provide just as much insight as a medical ethicist's professional opinion. The editors have two additional purposes in including these less known views. One, the editors encourage readers to respect others' opinions—even when not enhanced by professional credibility. It is only by reading or listening to and objectively evaluating others' ideas that one can determine whether they are worthy of consideration. Two, the inclusion of such viewpoints encourages the important critical thinking skill of objectively evaluating an author's credentials and bias. This evaluation will illuminate an author's reasons for taking a particular stance on an issue and will aid in readers' evaluation of the author's ideas.

It is our hope that these books will give readers a deeper understanding of the issues debated and an appreciation of the complexity of even seemingly simple issues when good and honest people disagree. This awareness is particularly important in a democratic society such as ours in which people enter into public debate to determine the common good. Those with whom one disagrees should not be regarded as enemies but rather as people whose views deserve careful examination and may shed light on one's own.

Thomas Jefferson once said that "difference of opinion leads to inquiry, and inquiry to truth." Jefferson, a broadly educated man, argued that "if a nation expects to be ignorant and free . . . it expects what never was and never will be." As individuals and as a nation, it is imperative that we consider the opinions of others and examine them with skill and discernment. The Opposing Viewpoints Series is intended to help readers achieve this goal.

David L. Bender and Bruno Leone,
Founders

Greenhaven Press anthologies primarily consist of previously published material taken from a variety of sources, including periodicals, books, scholarly journals, newspapers, government documents, and position papers from private and public organizations. These original sources are often edited for length and to ensure their accessibility for a young adult audience. The anthology editors also change the original titles of these works in order to clearly present the main thesis of each viewpoint and to explicitly indicate the opinion presented in the viewpoint. These alterations are made in consideration of both the reading and comprehension levels of a young adult audience. Every effort is made to ensure that Greenhaven Press accurately reflects the original intent of the authors included in this anthology.

Introduction

*"It is nonsense to talk about being 'for' or 'against'
technology itself. Technology is common to all cultures. It
is the act of turning anything into a tool, machine, or
procedure, with a practical end in view."*

— Lionel Basney, "Technolotry Unmasked,"
Other Side, *May/June 1997*

In 1965 Gordon Moore, cofounder of Intel, discovered that
the performance of a computer's memory chip doubled about
every eighteen months. Known as Moore's Law, his observa-
tion has proven to be remarkably accurate, as computing
power increased over 18,000 times from 1971 to 2000.

This explosion in computing power fueled the technological accomplishments that characterized the twentieth century. The invention of the microchip in 1958 spawned a generation of silicon-based technological wonders that have
recently become mainstream, including personal computers
and the Internet. In 1990 only 15 percent of households in
the United States owned a personal computer; by 1999 ownership had increased to close to 50 percent. In 1998, sales of
personal computers in the United States totaled $36 million,
and households with Internet access rose from 26.2 percent
in 1998 to 41.5 percent in 2000. The widespread use of personal computers and the Internet has provided previously
unfathomable conveniences to society.

As the world's fastest growing communications medium,
the Internet merges thousands of computer networks into
one international system. According to the Internet Society,
an international professional membership group that focuses
on issues surrounding the future of the Internet and Internet
infrastructure standards, the Internet is best described as a
"global network of networks enabling computers of all kinds
to directly and transparently communicate and share services throughout much of the world. Because the Internet is
enormously valuable, enabling capability for so many people
and organizations, it also constitutes a shared global re-

source of information, knowledge, and means of collaboration, and cooperation among countless diverse communities." The World Wide Web, the Internet's most popular system, combines business, government, personal, and educational "sites" and "pages" that present relevant text, images, and even audio and video data. Users may access information with the help of "browser" software, such as Internet Explorer and Netscape, and "search engines," such as Yahoo! and Excite. They may also contact other users through electronic mail (e-mail) and connect with others with similar interests using online discussion groups, bulletin boards, and "chat rooms."

By creating databases of educational resources for various businesses, libraries, nonprofit organizations, research institutions, branches of government, and others, the Internet has transformed the storage of and access to information. As stated by several leading computer scientists in their essay *A Brief History of the Internet*, "The Internet has revolutionized the computer and communications world like nothing before. The invention of the telegraph, telephone, radio, and computer set the stage for this unprecedented integration of capabilities. The Internet is at once a world-wide broadcasting capability, a mechanism for information dissemination, and a medium for collaboration and interaction between individuals and their computers without regard for geographic location." By opening up easily accessible pathways to a wide range of information, the Internet has created new opportunities for personal, educational, and business growth and has fostered the exchange of knowledge and customs of cultures around the world.

Many describe the Internet as a global "equalizer" that provides an unprecedented store of information to anyone with Internet access. These people claim that information technologies can provide the benefits of a large city to developing countries and rural communities around the world. Education may be brought to small villages that do not have teachers or classrooms via satellites and PC technology. The expertise of doctors can be instantly accessed thousands of miles from the actual location of the doctor. Many countries utilize "tele-medicine," or virtual appointments and examinations between doctors and patients. Finally, the information capability of the Internet is nearly unlimited, and the World Wide Web offers seekers information on virtually every subject imaginable. As stated by Michael C. Maibach, vice president of government affairs at Intel Corporation, "The Internet transmits digitized audio, video and data to any corner of the globe at any time. The people 'communicating' do not have to be on-line at the same time, nor share information in the same language. The Internet collapses space, time and language differences. The sources of data do not tire or misspeak. Information is limited not by accessibility but only by whether the information exists at all. In the Internet world, the word 'infinite' applies."

While most would agree with Maibach's assessment, some people have concerns about the effects of computers and the Internet on society. The most extreme of these critics are often referred to as Neo-Luddites. The original Luddites were a group of early nineteenth-century English rioters who waged war on technological advances in the textile industry, which they perceived as a threat to their way of life and livelihood. Neo-Luddites strive to break society's dependence on machines by rejecting technology and society's current ideology of progress. Neo-Luddites and other critics argue that despite the communication and information opportunities created by computer technology and the Internet, meaningful and fulfilling interpersonal relations have been replaced with relatively superficial e-mail and instant messaging. A study titled the *HomeNet* project, conducted by the Carnegie Mellon University in 1998, found that Internet use led to small but statistically significant increases in misery and loneliness and a decline in the overall psychological well-being of the participants. The project found that as people used the Internet more, they reported keeping up with fewer friends, spending less time talking with families, experiencing more daily stress, and feeling more lonely and depressed. Ironically, these results occurred even though interpersonal communication was their most important reason for using the Internet. Neo-Luddites and others argue that the avenues for communication opened by the Internet merely serve to connect people to machines rather than to other people.

Another concern is that technology and the Internet will increase inequality in society, as nearly 50 percent of Americans do not have personal computers in their homes. Minorities, the poor, and less educated citizens are the least likely to have computers; this disparity has been termed the "digital divide." A study released in 2000 by the Department of Commerce titled "Falling Through the Net" found that people with college degrees are eight times more likely to have a personal computer (PC) at home than those with only an elementary education. A high income household in an urban area is twenty times more likely to have Internet access than a rural, low-income household. Also, a child in a low-income white family is three times more likely to have Internet access at home than a child in a comparable black family and four times more likely than a child in a Hispanic household. According to the *Economist*, "Although Internet penetration has risen across all demographic groups, the digital divide remains only too real. It has also become a poignant proxy for almost every other kind of disadvantage and inequality in society."

Advances in computers and the growth of the Internet are among the incredible technological achievements of the twentieth century that have wrought significant changes upon society. While some consider such changes beneficial and embrace them, others, such as Neo-Luddites, perceive certain technological advances as threatening to personal relations and social dynamics. *Technology and Society: Opposing Viewpoints* examines several issues of contention in the following chapters: Has Technology Harmed Society? Are Technological Advances In Medicine Beneficial? How Has Technology Affected Privacy? How Will Technology Affect Society in the Future? Examination of these arguments will give readers a more thorough understanding of the impact of technological discoveries upon society.

CHAPTER 1

Has Technology Harmed Society?

Chapter Preface

The recent explosion in information technology has sparked controversy over what has been termed the "digital divide"—the alleged gap between those who have access to personal computers and the Internet and those who do not. While most people agree that some disparity exists between the haves and the have-nots, the debate lies in whether technology should be provided free of charge to poorer neighborhoods and schools.

Many people argue that the digital divide unfairly equips wealthy citizens and their children with more information and opportunities than those with lower incomes. According to Maureen Brown Yoder, associate professor of telecommunications, multimedia, and media literacy courses, "Until we enjoy universal access to technology, the Internet, and ideas on how to use them responsibly and productively, many people will wield an unfair advantage in their learning environment, in the job market, and in their daily lives." Yoder and others contend that free access to information technology should be available to all segments of society.

Others claim the digital divide is exaggerated, because even if poorer households do not own personal computers, Internet access is available at schools, public libraries, and community centers. Federal Communications Commission (FCC) chairman Michael Powell perceives technology as a luxury to be enjoyed by those able to afford it. He writes that "The term [digital divide] is dangerous in the sense that it suggests that the minute a new and innovative technology comes to market there is a divide unless it's equitably distributed among every part of the society. . . . I think there is a Mercedes divide. I'd like to have mine." Powell and others maintain that access to technology should not be federally subsidized.

Whether a problematic digital divide exists is one of several issues debated in the following chapter on the effects of technology on society.

> *"There is no future for a civilization that knows the price of everything and the value of nothing."*

Technology Has Harmed Society

Tom Mahon

Recent advances in technology have generated controversy over whether technological conveniences have benefited or harmed society. Many argue that innovations such as the Internet and cellular phones have contributed to a society proficient in processing data, but ignorant of humanity and culture. In the following viewpoint, Tom Mahon makes this argument, claiming that America fails to nurture spirituality and community in its race for technological superiority. Tom Mahon is a contributing editor to the *National Catholic Reporter*, a religious newsletter.

As you read, consider the following questions:
1. Why does the author consider technology to be "out of joint"?
2. What are the burdens passed on to the children of the "boomers," according to the author?
3. According to the author, what was the greatest legacy of the Enlightenment?

Reprinted, with permission, from "The Information Culture: Killing the Soul of the World," by Tom Mahon, *National Catholic Reporter*, February 14, 1997.

S omething has gone wrong with the Information Age. The microprocessor, the capstone product of our time, risks symbolizing technology run amuck.

There is a wide and growing perception that we no longer have a handle on technology, but rather get mis-handled by our technology. Twenty years ago we bragged, "Progress is our most important product." Now the young—the ones most comfortable with technology—wear T-shirts proclaiming, "Rage Against the Machine." Apparently the human spirit doesn't run under Windows.

Industry leaders eye new markets—"grab the consumers' eyeballs"—to provide even more intense entertainment to a public that is already over-entertained, overindulged. Meanwhile, during the latest Computer Dealer Exposition, the industrial world couldn't find a way to FedEx a can of Spam to Rwanda.

A single Pentium-based desktop computer has more processing power than NASA had at its disposal when it put men on the moon in 1969. Yet more than 90 percent of these machines today are used to turn memos and reports. An 8086 can do so much.

Confusion of Priorities

There is something profoundly out of joint when so much engineering ingenuity is frittered away while desperately needed for more pressing matters. Sixty percent of the world lives in abject poverty, but we don't see it because we desire more intense entertainment and less disturbing information. The sight of 50,000 children a day dying of starvation will not do a lot for the sale of Happy Meals.

We over-engineer the trivial or the deadly and are deaf to the cries of the many. We have the ingenuity to distribute 100 million land mines around the world, killing or maiming 500 people a week, but we can't seem to equitably distribute protein or vaccine globally in a global age.

The microprocessor might have been the great equalizer, capping off 250 years of "progress" in the wake of the Enlightenment and the scientific and industrial revolutions—that sweeping sequence of events that undercut the privileged, land-based aristocracies after 1750 and enabled and empowered more and more people.

But as we approach the end of the second millennium, it appears the information culture we are creating is leading us back to a two-tiered society with an increasingly small number of highly compensated information workers doing very well, even as more and more people—in the developed as well as the less developed world—are left behind.

This isn't to blame the microprocessor, which is morally neutral. That device only enables us; we ennoble or endanger ourselves with our tools.

Distinguishing Humanity from Technology

For all our information processing, we lack ability to get a human handle on technology. We teach literacy and mathematics in school, but we have no mechanism for infusing meaning and value into or out of our technology. We lack a means by which to "re-engineer" humanity into our tools. We have civil engineering but give no thought to civilization engineering.

" I AM NOW AVAILABLE BY PHONE, FAX, E-MAIL, SNAIL MAIL, VOICE MAIL, OVERNIGHT DELIVERY, CAR PHONE, CELL PHONE AND PAGER...........HIDE ME. "

Jim Borgman. Reprinted with special permission from King Features Syndicate.

We drown in information and are starved for meaning and knowledge. We are blinded by endless bit-level data and information but we lack any mechanism to infuse value-laden knowledge into the equation, to enlighten us and make sense of it all. And it is killing the soul of the world.

We "boomers" will write the valedictory of a millennium that dragged itself from serfdom to web surfing. But we send our children into a bitter new world. Burdened with a $4 trillion debt, poorly prepared by a struggling school system, living in a trashed environment and encultured by endless exposure to cynical, nihilistic "entertainment," with what optimism or enthusiasm will our young sound the keynote of their millennium?

The benefits we took for granted in this country—affordable housing, education, transportation, health care—are now increasingly reserved for the privileged, well-compensated, information-based elite, as was the case with the land-based aristocrats before 1750. In less developed countries (such as Somalia and Rwanda), the rule of law has simply broken down.

We have come so far with our technology in the last 250 years, raising the standard of living for many and the expectations of so many more. Yet now we seem to be returning to a two-tiered, feudal society with no secure middle in between. Perhaps the greatest legacy of the Enlightenment was the often-mocked middle class, whose existence kept the rich in check and the poor hopeful.

Potential Solutions

Permit a modest proposal for a way out of this impasse.

To retool and re-engineer our imbalanced industrial infrastructure, to equalize the unsustainable imbalance of global haves and have-nots and rebuild the infrastructure of the industrial world would be the greatest job-creation program in human history. Such useful human application of technology would consume far more microprocessors than Intel and AMD together could fabricate—not to mention being a monumental act of social justice.

And if justice isn't motivation enough, consider that it is impossible to build gated communities secure enough to keep out the growing rage and discontent.

There may be no room for nuns on the boards of technology companies, but there is no future for a civilization that knows the price of everything and the value of nothing.

We have done our children a great disservice with the debt we placed on them and the cynicism we foisted on them. We owe it to them—the first generation of the third millennium—to at least offer them a vision of a just and livable future. To show them realistic and concrete ways whereby we can use the information gathering and distribution capabilities of the silicon-based microprocessor—combined with the rich and literate tradition of justice and compassion from our carbon-based ancestors—to move with great urgency from bit level information processing to a more all-encompassing and humane knowledge processing and eventually, in God's good time, to wisdom processing.

| *"Fortified by the knowledge and power that come from the sciences, we may build on the finer values and wisdom of the ages."*

Technology Has Improved Society

V.V. Raman

In the following viewpoint, V.V. Raman, professor emeritus of physics at the Rochester Institute of Technology, describes the many scientific and technological breakthroughs of the twentieth century. While some argue that technological advances have lessened personal relations and community, Raman contends that science and technology have created avenues of communication and knowledge previously considered unattainable. He and others maintain that technology has improved access to information, medicine, and sanitation and continues to be an enormous benefit to society.

As you read, consider the following questions:

1. What three infectious diseases does the author claim medicine has learned to combat?
2. According to the author, why is harnessing nuclear fusion safer than tapping the electronic nucleus?
3. What is the "heartbeat" of today's computers, according to the author?

Excerpted from "Milestones of Twentieth-Century Science and Technology," by V.V. Raman. This article appeared in the May 2000 issue of and is reprinted with permission from *The World & I*, a publication of The Washington Times Corporation; copyright © 2000.

The twentieth century will be remembered for consciousness-raising and scientific/technological breakthroughs. This century made racism a shameful practice; recognized gender oppression as a social evil; proclaimed human rights as transcending race, caste, and religion; pleaded for international economic justice; began to celebrate diversity and to care for the disabled; and condemned exploitation of the young. It released millions from colonial shackles and established world organizations in which free nations join to solve problems of food and health, promote trade and education, and resolve political differences through discussion.

The twentieth century also made more scientific discoveries, introduced more technologies, and launched more assaults on the environment than all previous time spans combined. As one example, consider electricity: Through minibatteries and mammoth generators, from wind and waves, from sun and coal, energy is extracted to make electrons flow as the currents that light up the dark, heat the oven, and serve a hundred other needful or luxurious purposes. Humanity and electricity are forever bound together. And so it is with dozens of other profound contributions to science and technology. . . .

Telecommunication

From the moment speech began, human culture evolved. Indeed, society cannot continue without communication. Landmarks in communication have transformed civilization significantly.

Telegraphy, a child of the nineteenth century, was the first instance of telecommunication. From then on, telecommunication relied on advances in physics, especially electromagnetism. In 1876, Alexander Graham Bell transmitted his voice across eight miles over a wire: a first in human history. So began the saga of the telephone, which became essential equipment for the twentieth-century home, office, and factory.

In December 1901, Guglielmo Marconi sent radio signals across the Atlantic; a sound in England was heard right away in Newfoundland. News soon traveled fast and far by radio, and entertainment came into living rooms. By the mid-

1920s, inventors had managed to send images from place to place, initiating what would become TV, an invention with extraordinary potentials for informing, improving, and hurting society. Videotapes record events and sounds that can be experienced by generations yet unborn. Imagine how exciting it would be if we had videos of Socrates, Buddha, Caesar, and Christ!

Computers (1940s), artificial satellites (1950s), lasers and fiber optics (1960s) have all played a part in the telecommunication revolution. In the early 1980s, cellular phones were introduced in Chicago. Now they have spread the world over. Finally, an as yet unrealized dream of twentieth-century astronomers is to receive communication from intelligent life on a distant planet. What grander telecommunication could there be?

Antibiotics

Efforts to prevent and cure diseases are as ancient as civilization. Microorganisms, visible only under microscopes, were first seen in the 1600s. But the connection between diseases and minuscule creatures was not recognized until the latter half of the nineteenth century.

Conceptually, this discovery led to a simple solution for prevention and cure of diseases that could be traced to bacteria: kill the bacteria or inhibit their growth. During the nineteenth century, drugs such as quinine were already used against certain diseases. Many were discovered during the twentieth century, and a stupendous international infrastructure for manufacture and distribution of antibiotic pharmaceuticals was developed.

In 1915 Frederick Twort identified bacteria-eating viruses, or bacteriophages. Alexander Fleming discovered lysozyme in 1922: a bacteria-killing substance that our bodies produce. He also found that certain molds (penicillin) make lysozyme very effectively. This line of research proved very productive, generating a series of antibiotics such as streptomycin and erythromycin. In the 1930s, Gerhard Domagk and others discovered drugs like Prontosil and chemically synthesized molecules—called wonder drugs—which also destroy harmful bacteria. Rene Dubos (1930s) initiated techniques using microorganisms to produce antibacterial chemicals. In our century, medicine has learned to combat infectious diseases like pneumonia, tuberculosis, and typhoid, saving millions of lives.

Both overuse and underuse of antibiotics have spurred development of resistant bacterial strains. With tuberculosis killing more than two million people in 1998, staphylococcus infecting and killing patients in hospitals, and fatal pneumonia a possible outcome of severe colds, new chapters will, no doubt, be added to the story of antibiotics in the twenty-first century.

Nuclear Energy

Life is sustained by energy from the Sun, but what is the source of the Sun's endless energy? The twentieth century has found the process and replicated it.

Matter-to-energy conversions associated with atomic nuclei occur in varied forms and environments, yet conform to the formula $E = mc^2$, which gives the precise energy value of a given amount of matter. Thus, Sun and stars transform matter in their cores into radiant energy; nature on Earth has been releasing nuclear energy from radioactive substances since time immemorial. Scientists produced radioactivity in the 1930s and first harnessed nuclear energy from uranium by fission, in which heavy nuclei are split asunder. This rarely, if ever, happens spontaneously in nature.

Nuclear fission, used in the first atom bomb (July 1945), produced a blast equivalent to 20,000 tons of TNT. Controlled nuclear reactions power modern reactors. Submarines using reactors cruise for years without refueling. More than 430 nuclear power plants generate energy in many countries. Energy in the Sun and stars arises from the fusion of lighter nuclei (hydrogen and helium). We have replicated this too, for the first time in 1952 when a hydrogen bomb was detonated. To our knowledge, Earth is the only place outside of any star where nuclear fusion has occurred.

Once it was thought that tapping the atomic nucleus would answer all our energy needs, but serious problems loom. Aside from the nuclear arsenals of the world, which are tinderboxes for global annihilation, devastations of incalculable magnitude could result from reactor accidents. Then there is the problem of nuclear waste disposal. Burial deep underground in very thick storage tanks is one possibility. Harnessing nuclear fusion, which has proved very difficult, is a safer way of tapping nuclear energy, since there are few wastes here. This too may be accomplished in the twenty-first century.

Computers and the Internet

As with individual lives, human history is dramatically changed by unexpected events. The computer has transformed civilization. Initially designed as a computing machine, it soon became a device that could store, organize, manipulate, and retrieve vast amounts of information in incredibly short times. But computers are not just superefficient secretaries accessing superspacious filing cabinets. They not only think and follow commands but can make decisions, draw, design, scan labels, automate industries, calculate, translate, communicate, and more. Through the science of artificial intelligence, computers reveal how human minds may work, and some scientists think they will enable us to create replicas of the mind.

Power to the People

What all of [the] inventions [of the 20th century] so damned by other generations have in common is that they opened up new worlds to millions of people without discrimination as to wealth, power, or geography. These inventions make democracy viable in the 21st century. Today, the World Wide Web and Internet, for example, even with all of their indiscriminate information and trivia, are giving more people the opportunity to learn and discover than any invention in history. If knowledge is power, these inventions have given the power to most of the people—at least until those who control governments and businesses harness the inventions to their own profit.

Joe Saltzman, "Dick Tracy Never Had It This Good," *USA Today*, July 2000.

Early computers were made with vacuum tubes, which served radios and televisions of another era. Today their heartbeat is in the microchip, invented in the late 1950s and made possible by the transistor (late 1940s), which is based on discoveries resulting from quantum physics.

Microchips are found practically everywhere in modern society: planes, trains, cars, telephones, the water supply, offices, hospitals, the stock market, and schools. They have also created the Internet. Initiated for defense purposes in the late 1960s, the Internet has grown into a mammoth communication system linking countries and individuals across the world. It makes information on every topic accessible to anyone with a computer.

Late in the 1990s some feared that computers might fail and cause widespread chaos. This attitude perhaps symbolized their negative impacts and potentials: invasion of privacy, intrusion into military secrets, and sabotage. Computers have also created a multibillion-dollar industry, providing jobs to millions.

Perilous Passages Ahead

With all this, the twentieth century has also created stupendous problems, both pressing and potential. A population explosion in the face of diminishing oil reserves and farmable land, environmental pollution through automobiles and industrial effluents, perilous nuclear wastes, depletion of the rain forests: These are challenges of great magnitude. Then there are social and human problems, ranging from ethnic hatred and religious bigotry to poverty and malnutrition. So, though there is much to look forward to in terms of new technologies, increasing economic opportunities, interplanetary adventures, and possible cures for deadly diseases, we will be living in a fool's paradise if we are indifferent to the problems that will face mankind in the decades ahead.

The possibilities are immense and unpredictable, for the good and the bad: The discovery of a new and limitless non-polluting energy source could bring about a golden age of prosperity for all humanity. The rise to power of a mindless maniac with nuclear capabilities could unleash irrevocable devastation on our species. Education and science could free all mankind from ignorance and superstition, but resource scarcity could deepen the chasm between the haves and the have-nots. Religious and racial bigotry could fire simmering suspicions into horrendous conflagrations, or perhaps the emergence of an enlightened religious outlook would foster understanding and harmony among differing faiths. Or again, the long and checkered course of human history could be snuffed into a mere glitch in the planet's saga by the rude intrusion and blind fury of a stray asteroid lured by Earth's gravity. What awaits us in time, no one can tell. Not all the factors that shape the future are within our ken or control.

Recognizing these possibilities, let us join hands in our efforts to induce the positive and snub the negative potentials. Now, as never before in human history, we feel we are all passengers in the only spaceship we have. Fortified by the knowledge and power that come from the sciences, we may build on the finer values and wisdom of the ages and make our planet an even-more rewarding place to be.

| "Home access to the Internet is . . . the
nucleus of the digital divide."

Technology Has Created a Digital Divide

Mickey Revenaugh

Many claim that the revolution in information technology and the Internet has increased the gap between the haves and have-nots in society. The more wealthy families and communities are able to purchase computers and Internet access, which provides their children with more knowledge and opportunities than poorer children. In the following viewpoint, Mickey Revenaugh argues that the "digital divide" that this creates is unfair and contends that information technology should be available to all segments of society. Revenaugh is vice president of education for HiFusion, a free, filtered Internet service connecting kids, teens, schools, and homes.

As you read, consider the following questions:
1. What, according to the author, was the Technology Literacy Challenge of 1996?
2. According to the author, what does the broad term "community technology center" cover?
3. What does the author describe as the "red zone"?

Not long ago, I found myself at a Kinko's in Washington, D.C., not far from the Capitol in a neighborhood that—like a lot of D.C.—is mostly African American and struggling. I spend a lot of time at Kinko's all over the country, but this one was different. I was the only person there that the computer services clerk didn't know by name. Everyone else, hunched in front of a Mac or PC, or waiting by the self-serve copy machines to get on a computer, or pacing back and forth while the printer went through its paces, was clearly a regular. And everyone—save the sprinkling of folks working on resumes and fliers, but in the end, including them too—was at Kinko's to use the Internet.

At $12 an hour, that's expensive access. But when your need not to be left out of the Internet Revolution is intense enough, and you don't have a computer at home or at your job or at your neighbor's house, and the library's few machines are booked for hours a day, it's a price you apparently find a way to pay. Especially if it also buys you hands-on, nonjudgmental help from Mrs. Johnson at the computer services desk.

Despite the recent outcry from various quarters that access to technology should hardly be the first priority for improving the plight of the underserved, that D.C. Kinko's nevertheless brought home for me just how real the digital divide is—and also how complex and nuanced it is. . . .

Equity at School

If you look at it the right way, the entire history of education technology is a complex waltz with the issue of equity. Many of the earliest classroom computer advocates saw technology as a great equalizer, a powerful tool that could be deployed effectively by disadvantaged communities and families to close the learning gap. Assuring that technology resources were distributed equitably and used efficaciously has since proved to be quite the challenge for educators and policymakers alike, in part because advances in technology itself mean that new gaps open up as quickly as old ones are bridged. And since technology resources—from hardware and software to training and connectivity—are inevitably expensive resources, the challenge often comes down to finding ways to help those in greatest need foot the bill.

In the very beginning, of course, a kid's chance of having access to a computer anywhere was as random as a roll of the dice. You could never tell where you might find a teacher staying up all night to write BASIC programs for her lone classroom TRS-80—or having his class spend a month compiling an AppleWorks database of African American notables.

Bridging the Digital Divide

We must address the Digital Divide and make efforts to correct it, or the result will be serious social and economic splits, both nationally and internationally. We must narrow this gap to fully benefit from the Internet's enormous potential.

A U.S. Department of Commerce (2000) report clearly outlines the enormous need. As the number of jobs requiring technology skills grows, the disparity between those with access and those without—if not addressed—will "establish an impenetrable barrier not only to quality jobs, but also to educational opportunities and access to information that all Americans will need to be successful. The U.S. can avert a potentially devastating new social inequality between digitally literate 'haves' and 'have-nots' if the nation's skills, resources and commitment are mobilized quickly."

Maureen Brown Yoder, "The Digital Divide," *Learning & Leading with Technology*, February 2001.

But as technology began to catch on as an integral part of the school environment, divides opened up everywhere. By the time a struggling rural school installed its first Apple II lab, its suburban counterpart had upgraded to Macs. For every inner-city implementation of an integrated learning system focusing on basic skills, there was a countervailing deployment across town of multimedia PCs for student publishing and presentations. Computers multiplied, morphed, and became obsolete faster than anything that educators—accustomed to the lifecycle of textbooks—had seen. Teachers had to be trained, boxes needed to be connected in networks, and someone had to keep this whole new infrastructure up and running. Technology became an expensive, time-consuming imperative: one more measure by which have-not schools could fall behind.

But several assistive funding programs—many keyed to

low-income clients—began coming to the aid of schools and districts, making the idea of equal access seem like a true possibility. Federal Title I funds to help boost reading and math performance among disadvantaged children were channeled into technology by savvy school leaders; and in 1996, the Clinton administration laid down its Technology Literacy Challenge to provide access for all children to connect to computers with high-quality software and well-trained teachers. The Technology Literacy Challenge Fund and Technology Innovation Grants program began pouring millions into state and local technology efforts. The number of kids per computer—a key measure of technology implementation—began inching down, from 24 in 1989 to 12 in 1992 to just over seven in 1996. . . .

Access in the Community

The term "community technology center" is purposely broad, covering everything from stand-alone organizations to centers in housing projects (where the U.S. Department of Housing and Urban Development has funded a number of such centers), from public libraries to school-based programs, from totally local efforts to those associated with large national organizations such as the National Urban League and the Boys and Girls Clubs. And the clientele is equally diverse. In a recent study by CTCNet—an affiliation of 350 community technology centers around the country—the average age was 35; fully two-thirds were female; one-third each were white or African American, and 19 percent were Latino. Although 37 percent had not finished high school, 67 percent described themselves as students—pursuing everything from a GED to a college degree. One-third, excluding high school students and retirees, were unemployed. Three out of four reported family incomes of $30,000 or less. Perhaps surprisingly, 34 percent of those with incomes under $15,000 and 43 percent of those between $15,000 and $30,000 say they have computers at home; most use the centers because the technology there is newer and better and there's instruction on how to use it. . . .

The community technology center phenomenon is gaining steam, with infusions of dollars from the federal government, the private sector—including the PowerUP initiative supported by the Case Foundation, America Online, Gateway, Sun Microsystems, and others—and states that have seen major inflows of [government] funds and want to do more. One of these states is New Jersey, which has pooled federal funds with dollars from its own coffers to create the $7.4 million Access, Collaboration and Equity (ACE) Program, which will fund "after-hours" access to school instructional resources through community tech centers and school facilities. . . .

Not all community-focused energy, however, is going into access. The Center for Children and Technology (CCT) is drawing on its long history in "design for equity" to help centers like Access Harlem make certain the children of the neighborhood—particularly young girls—have something constructive to do online. Together with Libraries for the Future and an Australian-based software company Kahootz, CCT has designed Imagination Place!, a set of online animation tools and a "tinkering space" for kids to "create inventions the world forgot" says Dorothy Bennett of CCT. In addition to the Harlem center, Imagination Place! is being used in community technology centers in Newark, Hartford, Detroit, and Arizona, and is cited by the Children's Partnership study as an exemplar of online content that addresses the needs of the technologically underserved.

"Public access points, including libraries that have been wired and the whole array of community technology centers, are very much needed now and will be into the foreseeable future," says Jim McConaughey, senior economist at the National Telecommunications and Information Administration, the arm of the U.S. Department of Commerce that produces the Falling Through the Net reports. "Providing access to everyone at home is a worthy goal but it's doubtful that we'll ever be able to reach every household. We still have a significant number of people in this country who don't even have access to phones."

Home: The Final Frontier?

Home access to the Internet is, in some ways, the nucleus of the digital divide: If more people, and people from diverse circumstances, had Net connectivity at home, the crusade to wire schools and community centers wouldn't be so crucial.

As it is, according to Falling Through the Net:

- Rural black and inner-city Hispanic families are the least likely of any in America to own a home PC (18 percent and 21 percent respectively, compared to 47 percent of white households nationwide).
- Only 8 percent of inner-city female-headed households with kids under 18 have Internet access at home, compared to 28 percent of married couples with kids anywhere in the U.S.
- Over 16 percent of those who have a computer at home but don't access the Internet cite cost as a barrier (and another 8 percent say their computers aren't up to the task). . . .

Moving the needle on home access ever closer to the red zone—the point at which those remaining without Net access are only those who don't want it—is a goal shared by many in the technology world. Schools and community-based organizations are among those leading the charge.

School efforts to get a computer into every child's home are not new. In Indiana, for example, the statewide Buddy System Project has been placing PCs in the homes of large groups of elementary school students for the duration of one or more school years, and measuring the results. In addition to showing that kids with computers at home improve in critical-thinking, problem-solving, and research skills, Buddy dispelled notions that lower-income people are not interested in bringing technology home. "The Buddy Project and programs like it show that people are receptive, appreciative, and ready to take advantage of a home computer," says Saul Rockman, who's been involved with evaluation of the program. "In the poorest of homes, the computer usually sat by itself on a table in the middle of the house—like an altar. And as the kids learned, the parents learned."

Another persistent model is that of a laptop for every student. Microsoft and several hardware partners got the ball rolling in the U.S. with Anytime, Anywhere Learning, a school-based program in which each student has a portable PC for use at home and at school. The cost of ubiquitous laptops and the pressure it puts on schools to change their curriculum and logistics has meant that, until recently, implementation was limited to a relatively small number of schools (many of them nonpublic) and districts. Governor Angus King of Maine, however, put the laptop model back in the spotlight by proposing a $65 million program to give every seventh-grader in his state a laptop with Internet access. Meanwhile, companies like NetSchools have pushed ahead with a schoolwide technology model based on laptops for all, connected to each other and the Internet via infrared network during the day and equipped with modems for student and teacher use at home. . . .

Decreasing Costs

Efforts to get technology home with every kid are also being helped right now by market forces. The cost of PCs is continuing to drop, and there's resurgent talk about "network appliances"—inexpensive devices set up purely to access the Internet. Recent studies found that recent computer owners (those who purchased their computers within the past two years) are more likely than longer-term owners to be low income (30 percent versus 14 percent) and to have a high school education or less (59 percent versus 33 percent). In addition, there's a boomlet under way in free ISPs, which provide no-cost dial-up access in exchange for advertising space on the user's screen.

Of course, the challenges of universal home access are not just technological. "Too many of our Limited English Proficiency students don't have computers and Internet access at home, and that's only partly because when your family is struggling, a computer's not going to be at the top of the list," says Dade County educator Joanne Urrutia. "Many of our immigrant parents are afraid of technology. They're very protective of their children in this new culture, and they're not happy when they don't know what their kids are doing—which is the risk that the Internet presents." Urrutia believes that training for parents must accompany any effort to get technology home.

Children may provide some of the impetus toward breaking through home computer barriers. As recent studies report, kids without computers at home are far more likely than adults without computers to think that not having a computer at home is a problem, and kids are more likely than adults (37 percent to 17 percent) to feel left out because they don't have a computer.

With access beyond schools "on the front burner and nearing the boiling point," as policy analyst Norris Dickard puts it, we can expect to see continued attention on the home front from both policymakers and the private sector.

Beyond the Divide

Despite the somewhat mysterious ways in which race, income and education intertwine with Net access, and the steep challenge of reaching the hardest to reach, many observers are optimistic enough about closing the current digital divide that they're already looking beyond it. . . .

Kate Moore is also hopeful about seeing the closing of the digital divide. As president of the Schools and Libraries Division of the Universal Service Administrative Company, the organization that administers the E-rate, she's seen first-hand the progress that schools and communities are making to provide access for all—as well as what remains to be accomplished once that goal is reached. "What's important for the long term is 'technological literacy,'" she says. "Mastery of the tools of technology by everyone—whether for purposes of getting an education, or for success in the workplace—is critical to this nation's global competitiveness."

For researcher Reed Hundt, the challenge ahead is a conceptual one: "The next big thing will be to guarantee that all American citizens can participate in the Internet communities of their choice," he says. "That doesn't just mean the right to shop at Amazon.com. What is critical is that everyone should have the opportunity to use the Internet to access information relevant to their health; to be involved in their children's education; and to participate in their democratic government."

He's optimistic about that, as well: "Universal Internet access is a goal we absolutely can reach," Hundt says. "There are hundreds of alternative solutions, and all we have to do is find a way to develop them."

"The 'new new thing' in civil rights politics is just the latest variation on an old civil rights theme, the problem of inclusion."

Technology Has Not Created a Digital Divide

Eric Cohen

The "digital divide" refers to social inequalities many argue stem from unequal access to technological advances. In the following viewpoint, Eric Cohen, managing editor of the consumer advocacy journal *Public Interest,* contends that the alleged digital divide is the latest manifestation of the age-old civil rights argument against the exclusion of certain groups of people from society. Cohen and others argue that statistics belie the civil rights argument and that poverty and crime are more pressing threats to society than technology.

As you read, consider the following questions:
1. According to the author, why have civil rights and anti-poverty activists targeted companies?
2. How did the Commerce study exaggerate the gap between the "haves and have-nots," according to the author?
3. What does the author contend is the key factor separating technology users from nonusers?

Excerpted from "United We Surf," by Eric Cohen, *The Weekly Standard*, February 28, 2000. Copyright © News America Incorporated. Reprinted with the permission of *The Weekly Standard.*

O utside an August 1998 trade show in Santa Clara, Calif., a coalition of left-wing Bay Area groups denounced Silicon Valley for failing to share its wealth with minority consumers and employees. "Intel, Intel you're no good, / bring computers to the 'hood," the protesters chanted. An Intel spokesman complained to the *San Francisco Chronicle* that the giant chip-maker was being unfairly singled out; the company had a racially diverse workforce, she said, and besides, in the previous year Intel had donated $100 million in cash and equipment to education groups.

The spokesman had failed to grasp the essence of the fast-growing social justice movement that aims to end the "digital divide"—inequality between the rates at which rich and poor, black and white, use high-tech goods and the Internet. It's precisely because booming tech companies are progressive, charitable, and loaded with cash that the civil rights movement and anti-poverty activists have targeted them. And for the tech companies themselves—always keen on building market share—the idea of giving equipment to the poor is an attractive one, especially if their philanthropy is defrayed by government subsidies. The political class, for its part, . . . has proved highly enthusiastic. Few things can be more appealing to a politician than fighting for the poor by hobnobbing with billionaires. Perhaps unsurprisingly, given everything that's in it for them, the crusaders have barely slowed down to ask: Is there really a digital divide? Is it in fact a consequential social problem?

The New Trend

The digital divide is now the hottest social policy issue in Washington. It's the "new new thing" in civil rights politics. It has captured the imagination (and deep pockets) of major foundations, leading high-tech companies, the New Democrat economic-policy gurus, and even prominent Republicans—Virginia governor James Gilmore has challenged the high-tech community "to step forward and make a commitment to close the digital divide.". . .

In reality, the "new new thing" in civil rights politics is just the latest variation on an old civil rights theme, the problem of inclusion—or, in digital divide–speak, the problem of access. The argument is familiar: Blacks and Latinos (unlike Asians) have, on average, lower incomes than whites because they have been ignored by the old-boy networks, shut out of the capital markets, and excluded from the well-financed elite schools that make white people so wealthy. Institutional racism is still the norm, and new technologies only promise to exacerbate old divides. It's "technological segregation," says NAACP president Kweisi Mfume; "apartheid," says the Reverend Jesse Jackson. "Don't throw us aside, / close the digital divide," say the protesters in Silicon Valley.

The bible for this movement is a 1999 Commerce Department study—"Falling Through the Net: Defining the Digital Divide"—that activists cite with the agility of Talmudic scholars. "Whites are 2.5 times more likely to have home Internet access than Blacks and Latinos"; "the gap between whites and blacks grew by 53.3 percent between 1997 and 1998"; "more than a third of white families earning between $15,000 and $35,000 per year own computers, but only one-fifth of blacks do"—for reasons, [former] President Bill Clinton claims, that "we don't entirely understand."

There are other interesting statistics in the report: Asians at every income level are more likely than whites to own a computer; two-parent families of all ethnic groups are twice as likely to have Internet access as single-parent families (four times as likely among African Americans). But these statistics are not cited with the same frequency or alarm as the official statistics on the race gap. "There just aren't the advocacy groups in place for single-parents," says Anthony Wilhelm, director of communications policy at the Benton Foundation, perhaps the key player in the digital divide movement and a major beneficiary of AOL's multimillion-dollar largesse.

A Racial Ravine

On the subject of race, the official statistics tell an ambiguous story. The major piece of evidence for the Commerce Department's "racial ravine" is the following: Between 1994 and 1998, the gap between white computer ownership and black computer ownership grew from 16.8 to 23.4 percentage points. Sounds terrible. But read past the executive summary, and you discover the following: In 1994, 27.1 percent of white households and 10.3 percent of black households had computers. In 1998, the comparable figures were 46.6 percent for white households and 23.2 percent for blacks. Some basic arithmetic—conspicuously missing from the Commerce Department study, which presents the data in the most "alarming" possible way—shows that from 1994 to 1998, white ownership of computers rose 72 percent, black ownership rose 125 percent. In 1994, whites were 2.6 times as likely as blacks to own computers; in 1998, they were only twice as likely. The divide is not yawning wider; it's closing.

An Exaggerated Digital Divide

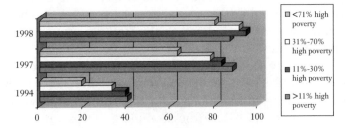

Percent of schools with Internet access

Testimony of Jason Bertsch, "The Role of Technology in America's Schools," Committee on the Education and the Workforce House Subcommittee on Early Childhood, Youth and Families, March 8, 2000.

This trend is consistent with another major study of Internet access—"The Digital Melting Pot," published by Forrester Research—which found that African Americans are getting home Internet access at a faster rate than any other ethnic group. Ekaterina Walsh, author of the Forrester study, projects that 40 percent of black households will be online at some point this year, while 44 percent of whites will—hardly a "racial ravine." Walsh gives three reasons for this: the rapid decline of computer prices; the increasing availability of free Internet access; and the surge of first-time computer buying during the 1998 and 1999 Christmas shopping seasons—periods not included in the Commerce Department study, which collected its data in December 1998. In Internet Time (computers are penetrating the market place seven times faster than electric service did and five times faster than telephones), December 1998 is another era. Even Larry Irving, former head of the National Telecommunications and Information Administration and the driving force behind the Commerce study, concedes that "we did miss a certain amount of information with regard to lower-priced PCs."

The Commerce study also exaggerated the "widening gap between technology haves and have-nots" by excluding computers outside the home—in the workplace, in schools and libraries—from its many white vs. black comparisons. Indeed, in its 1998 report, which is the basis for its many "alarming" comparisons, the Commerce Department did not even collect data on out-of-home access. Digital divide advocates skillfully blur the issue to their political advantage: If it weren't for E-Rate (the [federal government] program that uses new federal phone taxes to connect rural and inner-city schools to the Internet), they say, the digital divide would be worse. If government didn't step in, the racial ravine would be a racial abyss. But in their speeches, officials continue to use the Commerce Department figures that exclude school and work access—which is where most Americans, African Americans included, actually use the Internet.

As much as a crusade, closing the digital divide has become a cottage industry for many Washington–New York–Silicon Valley intellectuals, civil rights leaders, and philanthropy bureaucrats. The Rainbow/PUSH Coalition, the National Urban League, the NAACP, and the Leadership Conference on Civil Rights have all called the digital divide the "new frontier of the civil rights movement." The Commerce Department has created a digital divide clearinghouse—digitaldivide.gov—to monitor the nation's progress. Nine major corporations (AOL, AT&T, Bell Atlantic, BellSouth, Gateway, Intel, iVillage, Microsoft, SBC Communications), the Ford Foundation, and the National Urban League have partnered with the Benton Foundation to create the Digital Divide Network (DDN), a clearinghouse "to enable and facilitate the sharing of ideas, information and creative solutions."

Subjective Voices

But apparently not all ideas are worth sharing. Andy Carvin, senior associate at the Benton Foundation and editor of the DDN, told me that "the website is absolutely comprehensive." But the Forrester report and other critical articles—such as those by David Boaz of the Cato Institute and Adam Clayton Powell of the Freedom Forum—are nowhere to be found. Anthony Wilhelm, communications policy director at the Benton Foundation, says of the Forrester report and others that criticize the concept of a digital divide, "the values these reports promote are not appropriate for a democratic society."

B. Keith Fulton, director of technology programs at the National Urban League, is similarly dismissive: "The Forrester study was based on 1,500, maybe 2,500 people polled by telephone. A lot of poor people, maybe 20 to 30 percent, don't even have phones. The Commerce Department study went door-to-door to 48,000 people. Now who are you going to trust? Methodology becomes important here." Maybe so, but the Forrester report was in fact based on a mail survey of 85,000 people.

Not that either study is without weaknesses, but Forrester's numbers and projections are certainly more current and closer to reality than the government's claims of a widening "racial ravine." You get the feeling, though, that the digital divide movement has already moved well beyond the need to be grounded in fact. It's now grounded in the need to perpetuate a winning issue. If there is no divide, there is no movement, so there must be a divide. Or, as Wilhelm puts it, "The [Benton] Foundation's identity has become closely connected to the digital divide and other equity issues."

In fact, the Commerce Department study has some very interesting findings—two in particular—that are either not discussed or not effectively explained. The first is that blacks and whites with incomes over $75,000 per year own computers and use the Internet at roughly the same rate, while low-income whites are almost twice as likely to own PCs as low-income blacks. Second, the divergence between single-parent and two-parent households is striking: 61.8 percent of married couples with children own computers, while only 31.7 percent of female-headed households do. Dual-parent white families are twice as likely to have Internet access as single-parent white families; dual-parent black families are four times as likely to have Internet access as single-parent black families. An obvious reason, besides income, suggests itself: Men are more often early adopters of technology than women.

The Crisis of the Underclass

Altogether, the evidence suggests something like this: The economic boom of the last few years has made the vast majority of American families more wealthy; it has created millions of new high-paying jobs, especially in technology industries. As with white families, this has raised the incomes of millions of upwardly mobile black families, who now have enough money—and the desire—to buy computers. But there is a portion of the black community—a significant minority—that is not only chronically poor but burdened by unsafe streets, gang violence, and utter hopelessness. This group, not surprisingly, is not surfing the Web.

The key factor, as usual, is not race but income and marriage. In 1997, 69.2 percent of black children were born out of wedlock. This is the great tragedy that political leaders and captains of the computer industry who are philanthropically minded should be talking about. This—far more than "technological segregation" or "apartheid" corporate boards—is what shuts off poor children from American prosperity. Some sort of stable home life is what they need more than access to the Web. Which is, of course, the other great unsubstantiated claim of the digital divide movement: that what children especially need to succeed is more time in front of a computer. Skepticism about this claim actually grows the more one is familiar with how kids actually use computer access.

Couched in pro-market language and the hyperbole of the Internet age, the effort to close the digital divide is the latest version of the Jesse Jackson approach to social policy: talk about anything except the real cultural crisis of the underclass. To be sure, some of the digital divide efforts will have some positive effect—especially those dedicated to real mentorship rather than just computers in the classroom and technology courses for teachers. There are no doubt worse things big government and corporate America could be spending money on. But, on balance, this latest crusade— the "fourth movement in the civil rights symphony," Jackson calls it—is based more on myth than reality, and offers only mythical solutions to real problems.

"*Government has a responsibility to ensure that all its citizens . . . have access to technologies that can improve their lives.*"

The Government Should Regulate Technology

William J. Ray

In the following viewpoint, William J. Ray argues that the government has a responsibility to its citizens to provide the technological advances that will better their lives. The recent surge in information technology with personal computers and the Internet, he argues, has created an inequity within society that the government is responsible for rectifying. He and others contend that the government ought to regulate the prices and distribution of telecommunications technology, as private sectors unnecessarily inflate the costs and exclude less fortunate citizens. Ray is superintendent of Glasgow Electric Plant Board in Glasgow, Kentucky.

As you read, consider the following questions:

1. What was the first serious experiment with electricity, and when did it take place?
2. According to the author, how do investor-owned utility companies threaten democracy?
3. What effect, according to the author, does the Telecommunications Act of 1996 have on the FCC?

Reprinted, with permission, from "Power to the People," by William J. Ray, *Forum for Applied Research and Public Policy*, Winter 1998.

P ublic power is the result of a serendipitous arrangement formed more than 100 years ago. The most important ingredients were the human desire for a better life through the use of electric power and the philosophy of affirmative government—the idea that government has a responsibility to ensure that all its citizens, rich and poor alike, have access to technologies that can improve their lives.

A new technology—broadband telecommunications—could eclipse even electric power in its capacity to improve our lives. This system of fiber optic cables and electronics, capable of delivering competitive cable television, telephone service, and high-speed computer networking to every home and business, has the potential to invigorate local economies and improve the lives of citizens in the same way that electric utilities did in the first half of the 20th century. It needs only affirmative government to democratize it.

History Lesson

Contrary to popular belief, the development of electric power did not begin in the United States. The first serious experimentation on electricity took place in Europe in 1730 with the invention of the Leyden Jar, a device developed to build up and store an electric charge. But for decades the phenomenon remained a curiosity with no practical application. The real potential for electric power began to take shape during the early 1800s with the development of the electromagnet, the telegraph, and crude electric motors. America really became excited about electric power when early arc lighting was demonstrated in Paris and London in 1877. The prospect of lighting city streets launched the new industry.

From the very beginning, the visionaries creating the new industry were more interested in its profit potential than its potential ability to improve the lot of ordinary citizens. Because people wanted greater access to electricity than the industry was able or willing to provide, some cities sought, through affirmative government initiatives, to provide this service for themselves.

In April 1893, Detroit citizens voted 15,282 to 1,745 on an advisory ballot in favor of creating a municipally owned electric plant. Detroit Electric Light and Power (DELP) and its parent company, General Electric Company, fought against the effort. "If the city were to do its own lighting at about half what other companies bid, it would establish a bad precedent," warned William H. Fitzgerald, general manager of DELP.

The new mayor of Detroit, Hazen Pingree, a prosperous shoe manufacturer and municipal reformer, countered DELP's argument. "If this is done," Pingree declared, "it will take electric lights out of the luxuries of life, only to be used by the wealthy, and place it within the reach of the humblest of citizens." The resistance that Pingree encountered in providing the common people with electric power as a nonprofit service is the same that affirmative governments run up against today when they seek to democratize other technologies.

Successful Affirmative Government

For centuries, affirmative government has been used successfully in America as an economic development tool, but it has always been controversial. For example, when John Quincy Adams and Albert Gallatin, Thomas Jefferson's secretary of the Treasury, proposed a detailed 10-year plan for the construction of roads and canals by the government, their plan was rejected.

Gallatin contended that development of the transportation network could not "be left to individual exertion" because of its overarching importance to the future of the republic. Yet the system of improvements he and Adams proposed was never implemented. In 1837, Adams lamented:

> With this system ten years from this day the surface of the whole Union would have been checkered over with railroads and canals. It may still be done half a century later and with the limping gait of State legislature and private adventure. I would have done it in the administration of the affairs of the nation. . . . I fell and with me fell, I fear never to rise again, the system of internal improvement by means of national energies.

Fortunately, Adams was wrong. A hundred years later, his proposals were adopted by Franklin Roosevelt when he created the Tennessee Valley Authority (TVA) and, with it, public power. And in another 30 years, the federal government developed the interstate highway system.

Adams and Roosevelt both understood the power of affirmative government to advance economic development for the many instead of the few. They realized that turnpikes, canals, and electric systems need not always be built for the purpose of connecting thriving communities; rather, they could also be built to penetrate the wilderness in the hope of creating thriving communities. Roosevelt's strong belief in public power systems and his success in creating the TVA proved the wisdom of this philosophy.

New Threats

Public power systems have succeeded in democratizing electric power. Yet under the guise of deregulation, new initiatives by investor-owned utilities threaten to eliminate public power from the landscape. Investor-owned utilities—which first proposed our present system of regulation as a means of stemming the creation of public power systems—now want that system dismantled. They want to return to the good old days, before public power, when they could operate without competition.

The Internet and the New Economy

We are indeed at the dawn of a new age, and this age could well be better than the last. But to paraphrase Thomas Jefferson, eternal vigilance could well be the price of freeware. In the beginning, the primary allure of the Web was that everyone on it immediately had their own stage, their own printing press, and the government seemed out of earshot. Now that the Internet has become a backbone of corporate America and of the nation's thriving economy, it is getting more attention from the government than ever. [The Web] dearly needs it.

Frank Pellegrini, "Question of the Internet Age: To Regulate or Not to Regulate," *Time*, October 4, 2000.

They would even like to abolish regulations that prevent them from consolidating and expanding. Even now these utilities are merging and aggregating assets at a rate comparable to that of the 1920s, when they were in their heyday. The days of the trusts are returning.

How is this possible? Well, we have today a complete void of affirmative government philosophy. In 1912 even staunch private business proponents like Woodrow Wilson recognized that "without the watchful interference, the resolute interference of the government, there can be no fair play between individuals and such powerful institutions as the trusts." Yet in 1998, investor-owned electric utilities, telephone companies, and cable television companies have been able to convince ordinary citizens that Ronald Reagan was right: "Government is not the solution to our problem, government is the problem."

Threatening Democracy

Reagan's principles are still the rage in Washington. Unfortunately, his idea of getting government off the back of business has resulted in business on the back of government. While attacks on federal authority are conducted in the name of state and local rights, the real beneficiary is corporate power. As a result, the giant corporations threaten democracy itself.

Henry Adams examined this issue as early as 1870. The Erie Railroad, he wrote, had "proved itself able to override and trample on law, custom, decency, and every restraint known to society, without scruple, and as yet without check. The belief is common in America that the day is at hand when corporations far greater than Erie . . . will ultimately succeed in directing government itself."

Today, that belief is closer to reality than it has ever been. BellSouth, TCI, AT&T, and other corporations force our legislators to seek their approval before passing laws. A recent example is Microsoft's suggestion that our national economy would be slowed significantly if the government tried to interfere in its release of Windows 98. According to Bill Gates, what's good for Microsoft is good for America. It seems that Adams' pessimism was well founded.

New Possibilities

Like electric power in the early years of this century, broadband telecommunications offers rural Americans access to the same goods and services as their urban cousins. However, this technology needs the intervention of government to democratize its spread. The story of Glasgow, Kentucky, illustrates the benefits available to rural communities that practice the philosophy of affirmative government.

In 1988, the Glasgow Electric Plant Board (GEPB), a municipally owned public power system, responded to citizens' needs for competitive cable television and other telecommunications services by constructing a broadband telecommunications network throughout this rural community of 14,000 people.

The GEPB built a network capable of delivering services the private sector found unprofitable or just too risky. Its low-cost cable television service—providing 52-channel cable for just $14.25 per month—caused the incumbent private cable operator to lower its rates and improve its service. And the GEPB's four-megabit-per-second computer connection and Internet access costs only $20 per month. Comparable service from a telephone company might cost more than $1,000 a month.

The GEPB used its existing poles, trucks, billing system, and expertise to make broadband communications accessible to all Glasgow's citizens. Local residents have benefitted from the arrangement: lower cable rates have freed up millions of dollars to be spent in the local economy and created new jobs and new businesses. Furthermore, GEPB's high speed local area network (LAN) connections and Internet access have increased the profits and services of local businesses and helped them run more efficiently.

Improving Community

It is doubtful whether any private business would find the Glasgow broadband network financially attractive. While the network—which cost Glasgow residents roughly 4 million dollars to build—now produces enough revenue to make it a break-even proposition, it does not add to the city's coffers. Yet like other city infrastructure such as water and sewer connections, sidewalks, and parks, the network improves the life of Glasgow's citizens and encourages the growth of new businesses. And all these benefits have been achieved without increasing taxes or electricity rates.

GEPB's success could be replicated by other power companies around the country. But standing in the way are the same forces that Hazen Pingree had to face down in Detroit a hundred years ago. This time, the opposition comes not from General Electric, but from telephone and cable companies that convince state legislatures to ban affirmative government. These companies compel weak-minded state legislators to pass laws making municipal operation of broadband telecommunications systems illegal. By so doing, they condemn rural citizens either to live without the new technologies that have revolutionized business in urban communities or to pay exorbitant rates to behemoth cable and telephone companies.

The federal government also shares the blame here. The Federal Telecommunications Act of 1996 directs the Federal Communications Commission (FCC) to strike down any state or local law that has the effect of eliminating any entity from providing telecommunications services. Yet the presently impotent FCC refuses to do so for fear of stepping on the toes of state legislatures. Even though the federal government has affirmed the Bill of Rights against local vigilantism, preserved natural resources against local greed, civilized industry, secured the rights of labor organizations, improved income for the farmer, and provided a decent living for the old, today's FCC still feels powerless to implement the will of Congress. Had the present states' rights philosophy espoused by the FCC existed in the past, there would still be slavery in the United States.

Limitless Benefits

It is obvious that the democratization of electric power through affirmative government has provided, and continues to provide, unfathomable benefits to our people and our economy. The GEPB project allows us to glimpse the possible benefits of doing the same with broadband telecommunications. The time has come for affirmative government to set this process in motion.

The cable and telephone companies will surely howl about the socialistic implications of this course, but history does not validate their claims. Affirmative government programs such as public power have not set our nation on the road to total-itarianism. Fifty years ago, statesman Thurman Arnold scoffed at "the absurd idea that dictatorships are the result of a long series of small seizures of power on the part of the central government." In fact, the exact opposite is true. As Arnold pointed out, "every dictatorship which we know flowed into power like air into a vacuum because the central government, in the face of real difficulty, declined to exercise authority." Or, as Franklin Roosevelt observed, "History proves that dictatorships do not grow out of strong and successful governments, but out of weak and helpless ones."

The success of the mixture of electric power and affirmative government cannot be denied. Even though the investor-owned utilities have fought this mixture with every fiber of their being, in retrospect it has been good even for them. The profits of investor-owned utilities have steadily risen over the last 60 years, as has the standard of living in the United States. It is not time to water down this solution, but rather to apply it to the next great technological innovation: broadband telecommunications.

> "*Once inside the gates of the competitive free market, government is a Trojan horse.*"

The Government Should Not Regulate Technology

James K. Glassman

Technological advances have been privately funded by investors, which some argue favor the wealthy by creating unnecessarily high costs. Others contend that private enterprise and individual initiative have been the cornerstones of America's socioeconomic system, and that government regulation of technology would subvert capitalism and democracy in the United States. In the following viewpoint, James K. Glassman makes this argument and claims that government interference in technology and private business would be detrimental to the American economy. James K. Glassman is a fellow at the American Enterprise Institute and a writer and speaker on financial, economic, and political topics.

As you read, consider the following questions:

1. For what two reasons does Glassman advocate free market policies?
2. As paraphrased by the author, how does Alan Greenspan claim that wealth causes inflation?
3. What benefits does the author ascribe to tax-free commerce on the Internet?

Excerpted from "The Technology Revolution: Road to Freedom or Road to Serfdom?" by James K. Glassman, *Heritage Lectures*, no. 668, June 9, 2000. Reprinted with permission.

Let me just read a little bit from [Friedrich von Hayek's 1944 book] *The Road to Serfdom* on the subject of technology. John Blundell of the Institute for Economic Affairs in London was just telling me that this book was written during the Second World War. Hayek was an Austrian who had fled to England. He was in Cambridge, and was writing the book as he served as an air raid warden. He was a spotter, a fire warden, and he jotted down these thoughts in his spare time, writing one of the great books in history. He wrote:

> Well, it is true of course, that inventions have given us tremendous power. It is absurd to suggest that we must use this power to destroy our most precious inheritance, liberty. It does mean, however, that if we want to preserve it, we must guard it more jealously than ever, and we must be prepared to make sacrifices for it. There is nothing in modern technological developments that forces us toward comprehensive economic planning. There is a great deal in them that makes infinitely more dangerous the power a planning authority would possess.

And that's what bothers me. A planning authority at work on technology. My concern today is to help Americans and people all over the world to lead better lives through freedom. And technology is the engine for that improvement.

The Commerce Department reports that the communications, computer, and software industry accounted for an average of more than one-third of the growth in the economy from 1996 to 2000. Without this growth the U.S. economy would have increased at a rate of 2.6 percent, which is about average since World War II, instead of its actual rate of 4.3 percent. It's a huge difference. And falling prices in information technology have been lowering inflation by a full percentage point—meaning lower mortgage rates and lower costs for consumer loans.

But what can public policy do to continue and indeed to accelerate this prosperity? I spent 30 years as a journalist advocating free market public policies and principles for two reasons: first, because the prime value of this nation is human freedom. The Declaration of Independence says that we are endowed by our Creator with certain inalienable rights. Rights that can't be taken away. Among these are life, liberty, and the pursuit of happiness. And second, because free market principles produce prosperity. That's the lesson of the post–World War II era. But what are some of those principles?

Consumers First

One of the most important is a very simple one: Consumers first. This does not mean listening to self-appointed consumer advocates from Ralph Nader on down, who seem today to be in the thrall of plaintiffs' lawyers. No, it means allowing consumers to drive the market with their own free choices. Consumers naturally want more value at lower costs, and by this measurement, the computer industry has been the most consumer-driven industry in the history of the world. So how did this apparent miracle happen?

One thing we know is the government did not get between the creators of new technology and the consumers who stood to benefit from it. Nobody told the entrepreneurs in the garages of Silicon Valley and the garages of Redmond, Washington, what products to invent, how to sell them, what prices to charge, and which deals to offer to consumers—until lately that is.

Indeed, it is the largely unregulated nature of high technology that has produced the boom that we see today. The Internet has been a virtual regulation- and tax-free zone. It is a fabulous case study for the success of low intervention by government. But something is changing, and I believe it is fear of those changes that has helped to send the NASDAQ index down by about 30 percent in the course of a month. The new economy is threatening to begin to look more and more like the old—an environment in which the winners are not necessarily the companies that please customers the most, but the companies that do the best at keeping government at bay, or better yet, at *using* government to thwart competitors. Stock prices are falling because the risks to real innovators are rising.

Threats to Investors

I want to identify five threats that caused TechCentralStation to issue what we call an investor alert on April 3, 2000. I didn't say to investors that they ought to bail out of the market or abandon technology stocks. Instead, I said they should be vigilant about the new threats to their assets and their livelihoods. Even better, they should become active in opposing those threats. Otherwise, technology will prove to be not the road to freedom, but in Hayek's words, the "road to serfdom."

The Danger of Industrial Policy

The technology markets create enormous competitive opportunity, but great anxiety as well. Because of the pace of change and the quickness with which one can challenge established competitors, companies will naturally hunger for both regulatory assistance in inhibiting a competitor, and help in setting the direction of market developments (usually to the proponent's favor). Arguments to intervene often are couched as necessary to prevent anti-competitive behavior, even if such behavior has yet to occur. I strongly believe the Government has a role to play in stopping anti-competitive conduct, but it is a dicey business to do so based upon speculation about the future consequences in such a rapidly changing marketplace. Indeed, quite often, proponents of intervention are attempting to get the government to embrace a particular vision about the future and take actions in furtherance of that vision. This kind of activity used to go by the name of "industrial policy." I think such action is dangerous, because such a dynamic marketplace, driving by change and innovation, should not be tipped by government but should be allowed to play out.

Michael K. Powell, "Law in the Internet Age," speech before the D.C. Bar Association Computer and Telecommunications Law Section and the Federal Communications Bar Association, September 29, 1999.

What are those threats? First, what we like to call "the revenge of the middle-man." At a recent breakfast that I attended, an FCC commissioner noted that one of the joys of the Internet was that consumers are creating buying groups. By posting notices, 30 of them, for example, could get together and, as a more powerful force, get discounts if they all buy the same model of Chevrolet, for example.

Well, unfortunately in most states, these buyers can't get that Chevrolet directly from the manufacturer. It's against the law. Car dealers are busy in state legislatures getting laws passed that ban direct buying from manufacturers, even the buying of parts and insurance. Here's a real life example: When leases are up, auto manufacturers typically take back the used cars and sell them in closed auctions with dealers. Well, the automakers got an interesting idea: Why not post descriptions of the cars on the Internet and let consumers buy them online? Then, let each consumer choose a dealer for delivery of the car. Great idea. Saves money all around. Unfortunately, dealers in Texas put a halt to it since the process involved buying directly from manufacturers. That state shut down the Internet site.

In a sense, auto dealers are waging war against the American consumer, and so far the auto dealers are winning. Now dealers, suppliers, and other middlemen are feeling a squeeze in many industries because of the Internet. It's true. But rather than adapting to the new reality of high-tech commerce with what David Potruck of Charles Schwab calls "clicks and mortar" sensibility, they are appealing to politicians for help. And if this trend continues, such restrictions will slow the boom in Internet sales and derail this prime engine for growth.

Supply and Demand

The second threat comes from old-model, or "weird" model, monetary policy. Alan Greenspan's new theory is that wealth causes inflation. Here's what Greenspan laid out on March 6, 2000, in an important speech which I urge you to read. It's pretty appalling. I am paraphrasing: As technology helps make companies more productive and more profitable, their stock prices rise. Higher stock prices mean more household wealth. (So far, so good.) But that leads consumers to increase demand which pushes prices up, thus inflation, Greenspan concludes.

But I've got one question: What about supply? Higher stock prices make it easier for companies to raise capital which they plow back into their businesses, increasing supply, so prices don't have to rise since supply rises as demand rises. Look at the last ten years. Ten years ago household wealth and equities totaled about $2 trillion; today it's well over $10 trillion. But do we have more inflation than we had ten years ago? No. We have less. While Greenspan's stewardship has been largely productive, he now threatens to derail the surging economy, especially the capital-hungry high-tech sector, and we're watching it unfold right now.

The third threat is the increasing threat of companies running to politicians for help. Rather than battling for business in the marketplace, they seek advantages through government. For example, what does the consumer of high-tech products want right now more than anything else? If you spend any time on the Internet, you probably know the answer: speed. Unfortunately local telephone lines are still owned by government-created monopolies. And we've been waiting a long time for this to change. During an interview that we did at TechCentralStation, I asked Congressman Tom Bliley (R-VA), the chairman of the House Commerce Committee, why consumers were not getting broadband or fast access to the Internet. His answer: the local telecom bottleneck.

Promising Alternatives

The good news is that small companies are emerging and prodding these local monopolies. Companies like COVAD are offering high-speed service over phone lines. Satellite and wireless operators are developing their own competing services, and cable operators have now signed up more than one million customers for high-speed Internet connections. The technology is there, but unfortunately many of the Internet providers need the cooperation of local telecoms to get their service to customers, and they aren't getting it. Why?

In 1996 Congress and then President Bill Clinton agreed on a way to deregulate telecommunications for the benefit of consumers. It wasn't perfect, but it was the best solution possible. The idea was to let local telephone monopolies into the already competitive long distance business as long as the locals opened up their own markets to let other companies compete with them. For three years, the local Bell monopolies dragged their feet. Then, a little while ago, in New York, they were finally certified to open their markets enough to allow local companies to move into long distance.

But since then New York has been a disaster area. The technical job wasn't finished, certification was premature—and it should be a warning to other states and the Federal

Communications Commission (FCC). But even worse, recently local companies have convinced members of Congress to introduce a bill called H.R. 2420 that will let them into the data part of long distance, which will soon represent 80 percent of long-distance transmissions, without opening up as the Telecom Act of 1996 requires.

Now what will be the effect of the passage of that bill? Congressman Bliley said that it will *stall* the dissemination of broadband, and he's right; just when consumers need it most. And broadband is so important to [much of society].

New Ideas Can Come from Anywhere

The great thing about the Internet is that it is not mediated. We don't have Peter Jennings or the editors of the *Washington Post* in between the readers and the people who are telling the truth. In the battle of ideas, we win as long as our ideas are not mediated by somebody else. And that's what the Internet allows us to do—get the word out directly. But the Internet needs to be speeded up. It is absolutely urgent.

The reason that politicians should not get involved in technology is that by helping particular producers, they almost always dampen competition, which hurts consumers. The fact is that you never know where the next great idea is going to come from.

In the mid-1980s a husband and wife on the Stanford University faculty wanted to exchange e-mail love letters at work. Unfortunately, their academic departments used incompatible computer networks. So they created a new type of digital bridge over the divide, and in the process they created a new company. They named it Cisco Systems, one of the most successful businesses in American history.

As much as politicians may want to help, it's nearly impossible for them to know which companies will yield the greatest benefit to society. Who would have predicted that Cisco, now one of the three largest market-cap companies in the world, would have emerged from that e-mail love letter problem?

Still, sometimes in their zeal to support technology, politicians will be very tempted to help specific companies that say, "All we want is a level playing field." This usually means they want the government to give them someone else's property. And property rights are at the heart of technological progress. If companies can't be sure that, when they invest in a new product, they will own it, then they won't invest in it. And consumers will suffer.

Government Helps Companies, Not Consumers

Consider another example, also in the vital area of telecommunications. For a year America Online campaigned in Congress, in state legislatures, and in city councils to get laws passed that would force cable companies to permit AOL to use, at government-fixed prices, the cable pipelines that the cable companies, at great expense, laid down themselves. The cable companies, such as AT&T and Cox, have shown that they have every intention of selling access to their fast, fat pipes to content companies, but it is *they and their clients*, not city councils and legislatures, that must decide the terms. If you deprive cable companies of their property rights, why would they invest billions of dollars in high-speed access?

Then, in January 2000, AOL announced that it was buying Time-Warner, with 13 million cable subscribers, and suddenly the shoe was on the other foot. Now in an embarrassing reversal, AOL has said, in effect, "Well, never mind, we don't really want government intervention, at least not in our new-found business." But it may be too late.

I was recently in San Francisco, where the board of supervisors is moving ahead. Also, as a result of AOL's action, a federal court in Portland will soon rule on whether thousands of local governments can legally become Internet regulators. A lower court has said that they could. This decision would be disastrous for technology, and it shows the problem with inviting government's help. Once inside the gates of the competitive free market, government is a Trojan horse. . . .

Rising Internet Taxes

The fourth threat is no less damaging and coercive—the threat of rising Internet taxes. State legislators and local politicians see the Internet as a honey pot. In a short-sighted way they want to impose sales taxes on transactions across state lines, a move which studies show would deter e-commerce.

What's working like gangbusters in the e-economy is un-regulated commerce, free of sales taxes. This allows fast growth, which in turn is generating more wealth, higher off-line tax revenues, higher property tax revenues, and higher income tax revenues. Why would anyone want to disrupt that virtuous circle? . . .

Finally, threat number five: state attorneys general, in league with plaintiffs' lawyers, are setting their sights on rich New-Economy companies. The attorney general of Michigan was recently quoted as saying, "We want to do a Smith and Wesson–like thing with Doubleclick," referring to the gang-tackling tactic that politicians have used against gun manufacturers. Now the attorneys general are turning to a far richer target, high-tech companies. Doubleclick has been targeted for alleged privacy abuses, but it is only one in a series that includes eBay, the Internet auction firm, and, of course, Microsoft. Make no mistake: An all-out attack on tech companies is in the works, and it would be disastrous for the tech boom. . . .

Just as Hayek warned more than half-a-century ago that we were heading down the road to serfdom, I fear that government is now threatening us on the road to freedom, a road to a New Economy wrought by a new technology.

Don't let it happen.

Periodical Bibliography

The following articles have been selected to supplement the diverse views presented in this chapter. Addresses are provided for periodicals not indexed in the *Readers' Guide to Periodical Literature*, the *Alternative Press Index*, the *Social Sciences Index*, or the *Index to Legal Periodicals and Books*.

Jonathan Coleman	"Is Technology Making Us Intimate Strangers?" *Newsweek*, March 27, 2000.
James Fallows	"Technology We Hate," *American Prospect*, January 31, 2000.
John Gray	"The Myth of Progress," *New Statesman*, April 19, 1999.
Daniel S. Greenburg	"Delete the Revolution," *Lancet*, February 27, 1999.
Timothy C. Hoffman	"Finding the Blessing, Curses of Technology," *Computerworld*, August 14, 2000.
Herbert W. Lovelace	"Technology's New Manners—Notebooks and Cell Phones Are Changing Our Social Behavior—For Better or Worse," *InformationWeek*, March 27, 2000.
Peter G. Neumann	"Information Is a Double-Edged Sword," *Association for Computing Machinery*, July 1999.
Craig Peck, Larry Cuban, and Heather Kirkpatrick	"Schools Are Illsuited to Close the Digital Divide," *Los Angeles Times*, May 14, 2000.
Norman Podhoretz	"Science Hasn't Killed God," *Wall Street Journal*, December 30, 1999.
Ray Roth	"You Can't Shoot Down Technology," *High Volume Printing*, June 1999.
Kirkpatrick Sale	"We Love Technology Even as It Harms Us," *Los Angeles Times*, October 10, 1999.
Robert R. Selle	"What We Take for Granted," *World & I*, December 1999.
Royal Van Horn	"Technology: Violence and Video Games," *Phi Delta Kappan*, October 1999.
Thomas G. White	"The Establishment of Blame in the Aftermath of a Technological Disaster," *National Forum*, Winter 2001.
George K. Williams	"Let's Close the Racial Ravine on the Internet," *Computerworld*, August 16, 1999.

Are Technological Advances in Medicine Beneficial?

Chapter Preface

In an effort to eradicate such seemingly incurable diseases as Parkinson's, Alzheimer's, and some forms of cancer, scientists have begun to study human pluripotent stem cells, which are thought to be the primordial cells from which a human is created. Stem cells, derived from fertilized eggs just before they would have implanted in the uterus, have the power to develop into any type of cell in the body. Because the cells have the ability to divide indefinitely outside the human body without the signs of age that afflict other cells, scientists speculate that the cells may be able to grow tissue for human organ transplants and recreate genes to combat inherited diseases.

Stem cell research has caused much controversy because scientists obtain the cells from aborted fetuses and unused embryos created at in vitro fertilization clinics. Opponents of stem cell research maintain that the practice is open to unethical practices, such as the deliberate destruction of unborn children for research purposes. According to Congressman Bob Schaffer, "Killing preborn babies for tissue harvest is never justifiable. It is something no civilized nation should condone, much less fund with the tax dollars of conscientious, disapproving Americans."

Others contend that stem cell research is an invaluable tool for finding remedies for such degenerative ailments as heart and kidney disease. They argue that the embryos would never have become humans anyway, as the cells would have been destroyed through abortion or the destruction of the fetuses not used in the in vitro fertilization process. According to Arthur Caplan, an ethicist with the University of Pennsylvania, "We will not hold that person in a wheelchair hostage to our moral concerns about tissues that are going to be destroyed, [or] tissues that are not going to be turned into human beings under any circumstances, or cells of tissues [that may be misclassified as potential people]." Caplan and others maintain that allowing people to suffer because of ethical debates over tissue that would otherwise be destroyed is unfair.

Whether embryonic stem cell research is ethical is one of the issues debated in the following chapter on the technological advances in medicine.

"Medical advances would not occur without clinical trials."

Research Involving Human Subjects Is Vital to Medicine

Jennifer Brookes

The testing of new drugs and medical techniques on humans has generated much controversy. While some people allege that human subjects are exposed to unscrupulous practices and unacceptable risks, others maintain that human experimentation is invaluable to medical advancement, as many standard medical procedures evolved through clinical testing. Jennifer Brookes makes this argument in the following viewpoint. She contends that the many regulations on medical experimentation ensure the safety of the volunteers and society. Brookes is an editor and writer for *Closing the Gap*, the newsletter of the Department of Health and Human Services Office of Minority Health.

As you read, consider the following questions:
1. What are three types of clinical trials, according to the author?
2. According to the author, about how long does the drug development process take to complete?
3. What does an informed consent form include, according to Brookes?

Reprinted from "Clinical Trials: How They Work, Why We Need Them," by Jennifer Brookes, *Closing the Gap* (the newsletter of the Office of Minority Health, U.S. Department of Health and Human Services), December 1997/January 1998.

We've all taken medication. It may have been in the form of over-the-counter cough medicine, or prescription pills to treat chronic conditions such as diabetes. But how did researchers discover if the medicine is effective, if it's safe, and if there are any potential side effects?

Testing and evaluating drugs is serious business, and clinical trials are right at the center of the process, according to Dorothy Cirelli, chief of the Patient Recruitment and Referral Center at the National Institutes of Health (NIH), U.S. Department of Health and Human Services (HHS). "Medical advances would not occur without clinical trials," she said.

Well known as a world leader in medical research, NIH developed the first treatment for the human immunodeficiency virus (HIV), as well as innovative therapies for breast cancer, leukemia and lymphoma. For over a century, the agency has conducted clinical studies that explore the nature of illnesses. Clinical studies are currently underway for nearly every kind of cancer, HIV, cardiovascular disease, diabetes, obesity, and many other conditions—common and rare.

There are several types of clinical trials. By far the largest number are those that test new drugs. Prevention studies look at drugs or lifestyle changes that may help prevent disease. Diagnostic studies look at ways of detecting or finding out more about disease. Treatment studies may monitor new drugs or evaluate new combinations of established treatments.

The Food and Drug Administration (FDA), another HHS agency, is responsible for reviewing the scientific work of drug developers and implementing a rigorous drug approval process. FDA, which some have called the world's largest consumer protection agency, works to protect the public by ensuring that products are safe, effective, and labeled for their intended use.

From Animals to Humans

Every year, hundreds of clinical trials are conducted at medical centers across the country. Drugs must be studied in properly controlled trials in order to determine if they work for a specific purpose. Drugs that have not been previously used in humans must undergo preclinical test-tube analysis and/or animal studies involving at least two mammals to determine toxicity.

The toxicity information is then used to make risk/benefit assessments, and determine if the drug is acceptable for testing in humans.

NIH or the company developing the drug must conduct studies to show any interaction between the body and the drug. In addition, researchers must provide FDA with information on chemistry, manufacturing, and controls. This ensures the identity, purity, quality, and strength of both the active ingredient and the finished dosage form.

How Are Volunteers Protected?

The patient's rights and safety are protected in two important ways. First, the physician awarded a research grant by a pharmaceutical company or the National Institutes of Health must obtain approval to conduct the study from an Institutional Review Board. The review board, which is usually composed of physicians and lay people, is charged with examining the study's protocol to ensure that the patient's rights are protected, and that the study does not present an undue or unnecessary risk to the patient. Second, anyone participating in a clinical trial in the United States is required to sign an "informed consent" form. This form details the nature of the study, the risks involved, and what may happen to a patient in the study. The informed consent tells patients that they have a right to leave the study at any time.

Centerwatch: Clinical Trials Listing Service, "Background Information on Clinical Research," 2001.

The sponsor of the proposed new drug then develops a plan for testing the drug in humans. The plan is submitted to FDA with information on animal testing data, the composition of the drug, manufacturing data, qualifications of its study investigators, and assurances for the protection of the rights and safety of the people who will participate in the trial. This information forms what is known as the Investigational New Drug Application (IND).

The entire drug development process is lengthy and expensive. On average it takes about 10 years to complete. But it is an effective system that generally protects the public from dangerous and ineffective drugs.

The Three Phases of Clinical Trials

Treatment trials fall into three phases:

Phase I: These studies test the safety of a new treatment, providing humans with the first exposure to the drug. The studies help determine appropriate doses for further investigations. Trials usually involve a small number of volunteers.

Phase II: These studies assess how effective a treatment is. This phase continues to evaluate the safety profile of the drug and identifies the most common side effects. This phase usually involves a few hundred people. According to FDA, almost 80 percent of all drugs tested are abandoned by their sponsors after either the first or second phase because of drug toxicity or ineffectiveness. But if the results are promising, the sponsor moves to Phase III.

Phase III: This stage compares standard treatments with new ones. The trials, which primarily aim to obtain the necessary effectiveness data, may examine additional uses for a drug and consider additional population subsets. Much of the information obtained from this phase is used for the packaging insert and labeling of a medicine.

The safety and effectiveness data are usually audited by the FDA to verify the information submitted from clinical trials. The FDA also evaluates the manufacturing facility to ensure that a consistent, high-quality product can be produced. Then the drug is ready for marketing.

Understanding Benefits and Risks

The obvious benefit of clinical trials is finding new treatments and preventive measures. Another benefit is that participants of clinical trials have the opportunity to work with specialists who have the latest information on their disease. "In many cases research should be considered another treatment option," said Cirelli.

Phase III clinical trial participants receive either the most advanced and accepted treatment for a disease, known as the "standard" treatment. Or, they receive a new treatment that has shown significant promise for being as good as or better than the standard treatment.

But medical studies can carry unknown dangers and side effects. The risks and benefits of each study are explained in a document called an informed consent form that patients discuss with their doctors or nurses before agreeing to participate in the trials.

An informed consent form includes an explanation of the purpose of the study and expected duration; a description of any potential risks or discomforts to the patient; a list of benefits to the patient or others; a disclosure of alternative procedures or courses of treatment that may benefit the patient; information on confidentiality issues; and information on whether compensation is provided.

The process of informed consent is ongoing. A person entering a clinical trial continually receives new information about the treatment and progress. Signing a consent form does not bind anyone to the study; people can choose to drop out at any time.

Before any study begins, an institutional review board (IRB) must carefully review the plans for the study. Made up of scientists, doctors, clergy, and community advocates, the IRB aims to minimize risks to patients. This process ensures that any patient interested in the study can make an informed decision about participation.

"Whether the subjects are humans or animals, any assumption that experiments are always necessary, always carefully monitored, and always ethical is a fiction."

Research Involving Human Subjects Is Prone to Abuse

Neal D. Barnard

Advances in medical science such as new drugs or therapies undergo rigorous testing on animal and human subjects before they are offered to the general public. While many allege that animal and human experimentation is necessary to ensure the safety of new products, Neal D. Barnard argues that the inherent uncertainty and often unscrupulous procedures of such testing often endanger its subjects. Barnard maintains that the risks posed by human experimentation outweigh its potential benefits. Barnard is a physician and president of the Physicians Committee for Responsible Medicine, a nonprofit organization that promotes preventive medicine, broader access to health care, and higher ethical standards in medical research.

As you read, consider the following questions:
1. According to the author, how did human growth hormone become more readily available to the public?
2. What does the author claim are three negative consequences of human growth hormone?
3. Why does the growth of technology require more vigilance over experimentation, according to the author?

Reprinted, with permission, from "Human Experiments: Redrawing the Ethical Boundaries," by Neal D. Barnard, a fact sheet published on the website of the Physicians Committee for Responsible Medicine at www.pcrm.org/issues/Ethics_in_Human_Research/ethics_human_article.html. References listed in the original have been omitted in this reprint.

L ate 1993 was marked by revelations that hundreds of nonconsenting Americans had been used in radiation tests that began in the 1940s and continued much longer. The full facts of these experiments are not yet known. Earlier in the year, the National Academy of Sciences blew the lid off World War II chemical weapons experiments involving 60,000 American GIs, including at least 4,000 used in gas-chamber experiments that left many permanently disabled. For nearly 50 years, the victims kept their secret, having been told that if they revealed the military experiments, they could be charged with treason.

These examples and others like them—the Pentagon's hallucinogen experiments (1950–1975) and the Tuskegee syphilis experiments (1932–1972)—suggest that researchers can all too easily find themselves on the wrong side of ethical boundaries. It may be, however, that as scientific capabilities change, the boundaries themselves need to be redrawn. Consider some current cases.

Questionable Experiments Involving Children

At the National Institutes of Health (NIH), healthy short children are injected with a genetically engineered growth hormone. The 40 or so children already involved are not deficient in the hormone. They are simply short. Their parents consent to the experiment, and the children themselves affirm their assent. And 156 times every year until they reach their adult height, they get injections that are not medically necessary.

No one argues with the use of human growth hormone (hGH) for hormone-deficient children, who would otherwise be dwarfs. Until the early 1980s, they were the only ones eligible to receive it. Because it was harvested from human cadavers, supplies were extremely limited. But genetic engineering, beginning in the early 1980s, has changed that. The hormone can now be manufactured in massive quantities, leading pharmaceutical houses to eye a huge potential market.

Short stature, of course, is not a disease. The problems short children face relate only to how others react to their shortness and their own feelings about it. The hGH injections, on the other hand, may pose some risks, both physically and psychologically.

The injections speed the growth rate in 50 to 80 percent of nondeficient children over the short term. It is not clear, however, that final adult height is increased. There are indications that, for many children at least, the growth spurt simply occurs earlier. For those children who get no effect at all from the injections, the net result may be to aggravate feelings of shame and failure.

Growth hormone also causes the liver to manufacture more insulin-like growth factor, or IGF-1, which is thought to play a role in breast cell growth and lactation. It is not yet known whether children with more IGF-1 circulating in their blood have a greater risk of cancer or a poorer prognosis should cancer develop. However, several lines of investigation suggest this possibility. Test-tube studies show that IGF-1 encourages breast cancer cells to multiply, and it is even more potent in this regard than estrogens. Growth hormone may be one reason why women over 5'6" have double the risk of developing breast cancer than women below 5'3", particularly for premenopausal cancers. Tamoxifen, a drug used in the treatment of breast cancer, reduces IGF-1, an effect which may be partly responsible for its anticancer effect.

Growth hormone can also cause abnormal leanness, aggravate preexisting kidney disease, and stimulate the production of growth hormone antibodies.

The ethical questions raised by the experiment would not have to be asked, had technology not grown to the point of allowing the wholesale use of the hormone.

Control Groups Receive Inferior Treatment

Children are the subjects of other controversial NIH experiments. An $11.5 million pertussis vaccine trial in Italy included a placebo group—1,550 infants who received no protection against pertussis at all. The investigators expect that 5 percent of this group will develop the disease. An earlier NIH trial administered placebo vaccines to 2,600 Swedish infants. Such experiments would never be permitted in the United States, given the availability of an effective vaccine, and it was only after continued arm-twisting and money changing hands that the Europeans agreed to the NIH contract.

It should be noted that, in both the growth-hormone experiment and the vaccine trials, parents gave informed consent. However, parental consent does not remove ethical responsibilities in experiments on children. It is doubtful, for example, that a clinician prescribing anabolic steroids to children would be relieved of ethical problems simply because a parent consented to such treatments.

New ethical problems are also emerging in nutrition research. In the past, it was ethical for prevention trials to include a control group, which received very weak nutritional guidelines or no dietary intervention at all. However, that was before diet and lifestyle interventions, particularly those using very low fat, vegetarian diets, were shown to be able to reverse existing heart disease, push adult-onset diabetes into remission, lower blood pressure significantly, and reduce the risk of some forms of cancer. The Women's Health Initiative, finally under way after much political wrangling, includes a control group receiving much weaker nutritional guidelines. It may be that in the not-too-distant future, such comparison groups will no longer be permissible.

Unknown Dangers in New Drugs

A more widespread ethical problem, although one that has not yet received much attention, is raised by new pharmaceuticals. All new drugs are tested on human volunteers. There is, of course, no way that subjects can be fully apprised of the risks in advance, because that is what the tests are conducted to find out. Monetary compensation makes up for repeated blood tests and the other inconveniences that are routine for test subjects. But, should any serious health problem actually result, monetary compensation cannot begin to make up for the potential results. Manufacturers, of course, hope that animal tests will give a good indication of the potential risks. However, neither animal tests, nor the human premarket tests themselves, reveal the full range of drug risks. A U.S. General Accounting Office study found that of 198 new drugs entering the market between 1976 and 1985, 102 (52 percent) caused adverse reactions that premarket tests had failed to predict. And no less disconcerting, many of the drugs in question were unnecessary by any reasonable clinical standard. No fewer than eight were [sedatives, similar to Valium and Librium. Two were antidepressants, similar to others already on the market]. Several others were variations of cephalosporin antibiotics, antihypertensives, and fertility drugs. Certainly, some new drugs are necessary. But a great many new drugs are simply patentable variations of other successful drugs, sold to gain a share of a profitable market. The risks taken in this type of trial by subjects, and to a certain extent by consumers, are not in the name of science, but in the name of market share.

Human beings, of course, are not the only potential victims of unethical research practices. Given the emerging history of abuses and secrecy in human experimentation, the idea that animals—the 20 million chimpanzees, cats, dogs, and rabbits used every year in laboratories—will somehow be better treated is unconvincing, to say the least. Whether the subjects are humans or animals, any assumption that experiments are always necessary, always carefully monitored, and always ethical is a fiction.

Constant Vigilance Is Necessary

Ethical problems are not always, and probably not usually, the result of new technologies that have yet to be harnessed. There are always tremendous temptations for scientific investigations to go too far. Curiosity is a powerful human motivation which can lead well-meaning people to actions that are harmful, and even fatal, as a look at the most extreme cases clearly shows. When psychiatrist Robert Jay Lifton studied the experimenters responsible for the most hideous Nazi crimes, he found that, while some were clearly sadists, most were ordinary people in circumstances that permitted the full unfolding of human curiosity, propelling human aggression into the machinery of death.

The growth of technology only makes vigilance more necessary. Like growing human beings, growing science has strength that exceeds its control. And, like everyone else, scientists have to learn inhibition and restraint and can occasionally fail to inhibit an impulse of curiosity.

As governmental bodies review evidence of past abuses, the airing of buried secrets may improve vigilance against future abuses. But abuses will continue as long as there are experiments on subjects who are not in a position to give full informed consent and as long as technology provides novel ways of affecting their lives.

"*It is entirely feasible to have laws and professional practices that allow the giving or selling only of non-vital organs.*"

Selling Human Organs Is Ethical

The Lancet

In the following viewpoint, the editors of the *Lancet*, a journal of biomedical science, practice, and ethics, claim that the arguments against the sale of human organs are unreasonable and that the prohibition of such transactions is unfair. Although the *Lancet* focuses primarily on kidney donation, its arguments for a regulated system of compensated donation apply equally to other non-vital organs. The *Lancet* maintains that those opposed to the sale of body parts overlook the utter economic desperation that leads people to such extremes.

As you read, consider the following questions:

1. What do the authors claim is the "commonest" objection to kidney selling?
2. How do the authors refute the claim that organ donation must be altruistic?
3. How do the authors differentiate between public opinion and western public opinion?

Reprinted, with permission, from "The Case for Allowing Kidney Sales," *The Lancet*, vol. 352, no. 9120 (June 27, 1998), p. 1950. Copyright © 1998 Lancet Ltd.

When the practice of buying kidneys from live vendors first came to light some years ago, it aroused such horror that all professional associations denounced it and nearly all countries have now made it illegal. Such political and professional unanimity may seem to leave no room for further debate, but we nevertheless think it important to re-open the discussion.

The well-known shortage of kidneys for transplantation causes much suffering and death. Dialysis is a wretched experience for most patients, and is rationed in most places and simply unavailable to the majority of patients in most developing countries. Since most potential kidney vendors will never become unpaid donors, either during life or posthumously, the prohibition of sales must be presumed to exclude kidneys that would otherwise be available. It is therefore essential to make sure that there is adequate justification for the resulting harm.

Most people will recognise in themselves the feelings of outrage and disgust that led to an outright ban on kidney sales, and such feelings typically have a force that seems to their possessors to need no further justification. Nevertheless, if we are to deny treatment to the suffering and dying we need better reasons than our own feelings of disgust.

In this viewpoint we outline our reasons for thinking that the arguments commonly offered for prohibiting organ sales do not work, and therefore that the debate should be reopened. Here we consider only the selling of kidneys by living vendors, but our arguments have wider implications.

On Behalf of the Vendors

The commonest objection to kidney selling is expressed on behalf of the vendors: the exploited poor, who need to be protected against the greedy rich. However, the vendors are themselves anxious to sell, and see this practice as the best option open to them. The worse we think the selling of a kidney, therefore, the worse should seem the position of the vendors when that option is removed. Unless this appearance is illusory, the prohibition of sales does even more harm than first seemed, in harming vendors as well as recipients. To this argument it is replied that the vendors' apparent choice is not genuine. It is said that they are likely to be too uneducated to understand the risks, and that this precludes informed consent. It is also claimed that, since they are coerced by their economic circumstances, their consent cannot count as genuine.

Although both these arguments appeal to the importance of autonomous choice, they are quite different. The first claim is that the vendors are not competent to make a genuine choice within a given range of options. The second, by contrast, is that poverty has so restricted the range of options that organ selling has become the best, and therefore, in effect, that the range is too small. Once this distinction is drawn, it can be seen that neither argument works as a justification of prohibition.

If our ground for concern is that the range of choices is too small, we cannot improve matters by removing the best option that poverty has left, and making the range smaller still. To do so is to make subsequent choices, by this criterion, even less autonomous. The only way to improve matters is to lessen the poverty until organ selling no longer seems the best option; and if that could be achieved, prohibition would be irrelevant because nobody would want to sell.

The other line of argument may seem more promising, since ignorance does preclude informed consent. However, the likely ignorance of the subjects is not a reason for banning altogether a procedure for which consent is required. In other contexts, the value we place on autonomy leads us to insist on information and counselling, and that is what it should suggest in the case of organ selling as well. It may be said that this approach is impracticable, because the educational level of potential vendors is too limited to make explanation feasible, or because no system could reliably counteract the misinformation of nefarious middlemen and profiteering clinics. But even if we accepted that no possible vendor could be competent to consent, that would justify only putting the decision in the hands of competent guardians. To justify total prohibition it would also be necessary to show that organ selling must always be against the interests of potential vendors, and it is most unlikely that this would be done.

Acceptable Risk

The risk involved in nephrectomy [kidney donation] is not in itself high, and most people regard it as acceptable for living related donors. Since the procedure is, in principle, the same for vendors as for unpaid donors, any systematic difference between the worthwhileness of the risk for vendors and donors presumably lies on the other side of the calculation, in the expected benefit. Nevertheless the exchange of money cannot in itself turn an acceptable risk into an unacceptable one from the vendor's point of view. It depends entirely on what the money is wanted for.

Defending the Sale of Human Organs

The market in organs has its defenders. To refuse the sellers a chance to make the money they need, it is said, would be an unjustifiable form of paternalism. Moreover, the sellers may not be at greater risk living with one kidney, at least according to US research. A University of Minnesota transplant team compared seventy-eight kidney donors with their siblings twenty years or more after the surgery took place, and found no significant differences between them in health; indeed, risk-conscious insurance companies do not raise their rates for kidney donors. And why ban the sale of kidneys when the sale of other body parts, including semen, female eggs, hair, and blood, is allowed in many countries? The argument that these are renewable body parts is not persuasive if life without a kidney does not compromise health. Finally, transplant surgeons, nurses, and social workers, as well as transplant retrieval teams and the hospitals, are all paid for their work. Why should only the donor and the donor's family go without compensation?

David Rothman, *New York Review*, March 26, 1998.

In general, furthermore, the poorer a potential vendor, the more likely it is that the sale of a kidney will be worth whatever risk there is. If the rich are free to engage in dangerous sports for pleasure, or dangerous jobs for high pay, it is difficult to see why the poor who take the lesser risk of kidney selling for greater rewards—perhaps saving relatives' lives, or extricating themselves from poverty and debt—should be thought so misguided as to need saving from themselves.

It will be said that this does not take account of the reality of the vendors' circumstances: that risks are likely to be greater than for unpaid donors because poverty is detrimental to health, and vendors are often not given proper care. They may also be underpaid or cheated, or may waste their money through inexperience. However, once again, these arguments apply far more strongly to many other activities by which the poor try to earn money, and which we do not forbid. The best way to address such problems would be by regulation and perhaps a central purchasing system, to provide screening, counselling, reliable payment, insurance, and financial advice.

To this it will be replied that no system of screening and control could be complete, and that both vendors and recipients would always be at risk of exploitation and poor treatment. But all the evidence we have shows that there is much more scope for exploitation and abuse when a supply of desperately wanted goods is made illegal. It is, furthermore, not clear why it should be thought harder to police a legal trade than the present complete ban.

The Lesser of Two Evils

Furthermore, even if vendors and recipients would always be at risk of exploitation, that does not alter the fact that if they choose this option, all alternatives must seem worse to them. Trying to end exploitation by prohibition is rather like ending slum dwelling by bulldozing slums: it ends the evil in that form, but only by making things worse for the victims. If we want to protect the exploited, we can do it only by removing the poverty that makes them vulnerable, or, failing that, by controlling the trade.

Another familiar objection is that it is unfair for the rich to have privileges not available to the poor. This argument, however, is irrelevant to the issue of organ selling as such. If organ selling is wrong for this reason, so are all benefits available to the rich, including all private medicine, and, for that matter, all public provision of medicine in rich countries (including transplantation of donated organs) that is unavailable in poor ones. Furthermore, all purchasing could be done by a central organisation responsible for fair distribution.

It is frequently asserted that organ donation must be altruistic to be acceptable, and that this rules out payment. However, there are two problems with this claim. First, altruism does not distinguish donors from vendors. If a father who saves his daughter's life by giving her a kidney is altruistic, it is difficult to see why his selling a kidney to pay for some other operation to save her life should be thought less so. Second, nobody believes in general that unless some useful action is altruistic it is better to forbid it altogether.

It is said that the practice would undermine confidence in the medical profession, because of the association of doctors with money-making practices. That, however, would be a reason for objecting to all private practice; and in this case the objection could easily be met by the separation of purchasing and treatment. There could, for instance, be independent trusts to fix charges and handle accounts, as well as to ensure fair play and high standards. It is alleged that allowing the trade would lessen the supply of donated cadaveric kidneys. But although some possible donors might decide to sell instead, their organs would be available, so there would be no loss in the total. And in the meantime, many people will agree to sell who would not otherwise donate.

It is said that in parts of the world where women and children are essentially chattels there would be a danger of their being coerced into becoming vendors. This argument, however, would work as strongly against unpaid living kidney donation, and even more strongly against many far more harmful practices which do not attract calls for their prohibition. Again, regulation would provide the most reliable means of protection.

The Danger of the Slippery Slope

It is said that selling kidneys would set us on a slippery slope to selling vital organs such as hearts. But that argument would apply equally to the case of the unpaid kidney donation, and nobody is afraid that that will result in the donation of hearts. It is entirely feasible to have laws and professional practices that allow the giving or selling only of non-vital organs. Another objection is that allowing organ sales is impossible because it would outrage public opinion.

But this claim is about western public opinion: in many potential vendor communities, organ selling is more acceptable than cadaveric donation, and this argument amounts to a claim that other people should follow western cultural preferences rather than their own. There is, anyway, evidence that the western public is far less opposed to the idea, than are medical and political professionals.

It must be stressed that we are not arguing for the positive conclusion that organ sales must always be acceptable, let alone that there should be an unfettered market. Our claim is only that none of the familiar arguments against organ selling works, and this allows for the possibility that better arguments may yet be found.

Nevertheless, we claim that the burden of proof remains against the defenders of prohibition, and that until good arguments appear, the presumption must be that the trade should be regulated rather than banned altogether. Furthermore, even when there are good objections at particular times or in particular places, that should be regarded as a reason for trying to remove the objections, rather than as an excuse for permanent prohibition.

The weakness of the familiar arguments suggests that they are attempts to justify the deep feelings of repugnance which are the real driving force of prohibition, and feelings of repugnance among the rich and healthy, no matter how strongly felt, cannot justify removing the only hope of the destitute and dying. This is why we conclude that the issue should be considered again, and with scrupulous impartiality.

"[An] ethical difficulty with an organ market is that it commodifies body parts and alienates part of a person against that person as a whole."

Selling Human Organs Is Unethical

William E. Stempsey

The shortage of viable human organs for people who need transplants has created a lucrative market in such countries as India and Egypt. While the practice remains illegal in the United States, controversy has arisen over whether an individual ought to have the right to profit from his or her own organs. In the following viewpoint, William E. Stempsey claims that the system of selling organs is so riddled with greed and corruption that desperately poor sellers risk being manipulated by wealthier buyers. Stempsey is a medical doctor who works as an assistant professor of philosophy at the College of the Holy Cross and serves on several ethics committees.

As you read, consider the following questions:
1. According to the author, what are the three characteristics of an anonymous act?
2. What does the author mean when he claims that to sell a body part is "to alienate an essential part of the human person against another"?
3. As quoted by the author, how does John Ruskin differentiate between wealth and "illth"?

Excerpted from "Paying People to Give Up Their Organs: The Problem with Commodification of Body Parts," by William E. Stempsey, *Medical Humanities Review*, Fall 1996. Reprinted by permission of the Institute for Medical Humanities, University of Texas Medical Branch, Galveston, Texas.

The sale of human organs remains illegal in the United States. The National Organ Transplant Act of 1984 makes it unlawful for any person to "acquire, receive, or otherwise transfer any human organ for valuable consideration for use in human transplantation if the transfer affects interstate commerce." A shortage of transplantable organs remains a problem in the United States, and throughout the world as well. Many proposals, such as changing the definition of brain death from "whole brain" to neocortical brain death, have been offered as ways to provide greater opportunities for organ harvesting. For more than a decade, some have suggested that commercializing the organ procurement system would help to alleviate the shortage of transplantable organs by serving as an incentive for people to "donate" organs. Some also suggest that paying for organs would further the cause of justice for living donors by properly compensating them. Jack Kevorkian calls purely altruistic organ donation a "delusion of hypocrisy." He sees unfairness in prohibiting people from selling their organs, as surgeons are rewarded handsomely for their efforts in removing and transplanting organs while those who are the source of the organs are expected to behave altruistically.

Is our prohibition of the selling of human organs merely a sentimental custom that ought to be abandoned for the sake of increasing the supply of transplantable organs and eliminating the tragedy of the deaths of people waiting for transplants? Or does this prohibition reflect a truly "tragic choice" that plagues us because of the conflict of values, both of which ought to be maintained?

Besides those that call for an open market for transplantable organs, there have been many sophisticated proposals for compensating organ donors. Law professor James Blumstein, for example, argues for allowing "forward contracts for transplantable cadaveric organs." This would provide payment for a person who would not give up an organ until death. Variations on this proposal might include tax credits or paying for burial expenses for one who gives up an organ at death. These sorts of proposals deserve further debate. However, to increase the number of available organs, any incentive system would have to induce a high percentage of present nongivers to give while alienating very few of the people presently willing to give. In addition, providing incentives would drive up the cost of transplants because we would then be paying for organs that are now freely given.

My concern, however, lies in a more foundational aspect of organ procurement: the proper valuation of human beings and the organs that are parts of them. To highlight this aspect, I will circumnavigate some of the complexities of proposals such as those of Blumstein and will limit my discussion to the issue of paying living donors to give up their organs. Although this practice is illegal in the United States, it is not universally prohibited. In some places, Egypt for example, selling organs is not illegal. Kidneys there may sell for as much as $10,000 to $15,000. Desperately poor young men are drawn to dilapidated boarding houses in Cairo where they may wait for several months to be matched with a well-to-do person needing a transplant. My arguments will apply as much to places like this as to proposals to allow an organ market in the United States. . . .

The Importance of Informed Consent

Even those who favor allowing people to sell their organs highly value the principle of informed consent and would abhor any coercion directed toward those making such decisions. Informed consent is of paramount importance in any scheme of justice. It is of paramount importance in the surrendering of body parts, even when commercial interests are not involved.

Informed consent has usually been seen as a matter of exercising personal liberty or autonomy. A person acts autonomously only if he or she acts (1) intentionally, (2) with understanding, and (3) without controlling influences. One might sell one's organs intentionally. Understanding, however, might be deficient if full information about the consequences of living without a particular organ or the risks of surgery for organ removal are not fully explained. Thus, even those who favor allowing the selling of organs agree that full information should be given and understood before an organ is sold. The crux of the matter of informed consent in selling an organ is whether there has been controlling influence in the decision.

The Three Types of Controlling Influence

Controlling influence can take the form of coercion, persuasion, or manipulation. Coercion involves controlling a person's behavior by the negative sanction of a threat. The threat must be severe, credible, and irresistible. Offering a poor person the possibility of making money by selling an organ does not constitute coercion in this sense because it is not a threat. Coercion is involved only if some harm were threatened to a person who refused to sell the organ.

Persuasion is the inducement, through appeals to reason, freely to accept the action advocated by the persuader. This constitutes no problem for informed consent, for persuasion aims only at influencing the free choice of another. A person may be persuaded to adopt a particular point of view, but as long as that happens freely, the ability to give informed consent is not altered.

Doctors Against Selling Organs

The predominant ethical issue in transplantation over the last few years has been the attempt in some Western countries to develop a market in organs. In a free market economy it is very tempting to consider a transplant organ as a commodity. Some economists certainly espouse this view, and, in doing so, come into a head-on collision course with the traditional medical view that an organ transplant is a priceless gift to the recipient and organ donation should always be an altruistic act. The medical view goes on to state strongly that organs should not be bought or sold, either directly or within the framework of a futures market; in this, the doctors in the West are supported by the World Health Organization, which has ruled against physicians being involved in transplantation if they have any reason to believe that the organs have been the subject of commercial transactions.

Robert A. Sells, *Birth to Death: Science and Bioethics.* New York: Cambridge University Press, 1996.

The dubious possibility of giving true informed consent to sell one's organs is explained by the concept of manipulation. Manipulation is influence that noncoercively alters the actual choices available to a person or nonpersuasively alters the person's perception of those choices. There are three classes of manipulation: manipulation of options, manipulation of information, and psychological manipulation. The poor person who resorts to selling an organ to better his or her life in some material way is in a situation of desperate need. In such situations, even in the absence of manipulation of information, manipulation of options and psychological manipulation are very likely.

The extent to which unwelcome offers of options compromise ability to give informed consent depends on the ease or difficulty the person being manipulated finds in attempting to resist the offered option. Rewards of "mere goods" are generally easier to resist than rewards that stand to alleviate harms; so, the former are generally more compatible with autonomous action than are the latter. Those who argue that allowing poor people to sell organs would benefit them by providing needed money tend to see the offer of money for organs as a "mere good." The motivation of most potential organ sellers, however, is the desire to alleviate their situation of desperate need. The offer of money to someone who is desperately poor makes any means of getting the money, including selling organs, hard to resist and therefore manipulative.

Those soliciting organs are liable to employ psychological manipulation as well, even if unwittingly. Information that a seller might find important could easily be paternalistically judged by the solicitor as unimportant. It would be too easy to take advantage of a desperately poor person by withholding information about medical risks of organ removal or the difficulties of receiving follow-up medical care in the case of complication until after a choice had already been made. . . .

Alienating the Body

A second ethical difficulty with an organ market is that it commodifies body parts and alienates part of a person against that person as a whole. Almost everyone highly values principles such as "inviolability and inalienability of the human person." Such general principles can admit a range of specific content, however. Certainly such principles must apply to a person's body and not only to some theoretical aspect of the person such as autonomy. If the body were considered to be property, it could be bought and controlled by others; this would amount to slavery.

Some, however, find it hypocritical that hospitals, physicians, drug companies, and the medical-industrial complex in general should become enriched through organ transplantation and the commercial development of human biological material while the donor of the tissue is expected to be purely altruistic. They see the granting of property rights to one's body as a way of safeguarding the same values of inviolability and inalienability.

A free market in organs presumes that we own our organs and have a right to sell them. But does it make sense to talk about ownership of one's organs, or of one's body? Historically, the common-law tradition has viewed body materials as having no value in themselves. So, organs were considered to be legally abandoned on severance unless the person from whom the organs came took timely steps to direct their disposition.

To hold organs as property, one must assume a Cartesian mind-body duality and, further, that the essence of the human person is the mind, the body being a subject of the mind's reign. However, human beings as we know them on this earth are essentially embodied beings. That is, we do not recognize a human being without a body. Having a mind and, many would argue, a spirit is also an essential part of being human. If this is true, then the human person is essentially a combination of body, mind, and spirit, and not merely a mind in possession of a body. Thus, to assert the dominance of mind over body, which must be done to justify selling a part of one's body, is to alienate one essential part of the human person against another. . . .

Favoring the Wealthy

In this section I want to explore a third way in which a commercial organ market would exploit sellers of organs. A system is exploitative if the benefits of transactions carried out according to the accepted procedures of the system predictably go to the more well-off at the expense of the less well-off. It seems inevitable that in a commercial system those who are least well-off in terms of financial resources would be the ones to sell their organs for the benefit of those who can afford to pay for transplants or health insurance adequate to cover transplants. A crucial question, then, is whether selling an organ can leave one better off. . . .

In India, the kidney market has been dubbed with the euphemism "unconventional renal transplantation." Even though corruption and bad results are prevalent in many Indian centers using paid donors, defenders of commercialism say, "It is people who are at fault, not the system." The nature of the system, however, is that it is composed of people from vastly disparate economic classes: the buyers comprise those who can afford the considerable expenses of buying an organ and having it transplanted, and the sellers, as in Egypt, are usually the desperately poor. Often, only a small part of what a desperate recipient is willing to pay is received by the seller: the rest goes to brokers who sometimes are part of the medical team. Some Middle Eastern patients who received transplants in India report that less than ten percent of what they paid for their transplant went to the organ seller. In addition, sellers, once their organs are removed and they are released from the hospital, have no protection and usually cannot afford the costs of hospitalization should postoperative complications develop.

The idea of trading for mutual advantage provides a rationale for organ-market advocates. There is, however, reason to question whether mutual advantage is served in an organ market. In coming to agree on a price for an organ, each party seeks advantage for itself. But bargaining advantage will almost always go to a wealthy purchaser over a poor seller. This advantage, which is not only financial but also social and educational, makes the two sides so unequal as to bring into question the fairness of the transaction. Justice turns out to be nothing but . . . the advantage of the stronger.

There is no way adequately to regulate prices to avoid this sort of exploitation. If, on the one hand, some fixed dollar amount were set as payment for organs, the payment would be a relatively high percentage of a poor person's net worth and so would be a powerful inducement for the poor person to sell the organ. The same payment would be but a small percentage of a rich person's net worth and so would be an insufficient incentive for a rich person to sell an organ. If, on the other hand, payments were regulated according to some percentage of the seller's net worth, then the result would be a system in which poor sellers made very little from the sale, while the rich would get richer. The incentive to sell would apply only to the rich, who, because they are not in financial need, would be unlikely to want to sell an organ. If payments were set according to some percentage of the buyer's net worth, then sellers would seek buyers who could afford to pay the highest prices. This would be least exploitative for the sellers, but would yield the unacceptable result that organs would be distributed according to ability to pay. . . .

Wealth or Illth?

Finally, it will be good to consider just what constitutes wealth in the context of organ transplantation. Earlier we raised the question of whether selling an organ can leave one better off. New light is shed on this question by an old idea prevalent in mid-Victorian literature: "a widespread insistence that economic value can only be determined in close relationship to bodily well-being." In *Unto This Last*, author John Ruskin relates an incident of a ship sinking off the coast of California in the gold rush days. One of the passengers had fastened around himself a belt with two hundred pounds of gold in it, with which he was found afterwards at the bottom. Ruskin asks: "Now, as he was sinking—had he the gold? or had the gold him?" Ruskin concludes that

> possession, or "having," is not an absolute, but a gradated, power; and consists not only in the quantity or nature of the thing possessed, but also (and in greater degree) in its suitableness to the person possessing it, and in his vital power to use it.

Wealth, then, is the possession of articles we can use. We ought to have a corresponding term such as "illth," says Ruskin, for possessions that cause "Various devastation and trouble around them in all directions."

Might we apply these conceptions of wealth and illth in talking about selling human body parts? To someone who needs an organ transplant, a part of someone else's body might indeed be worth its weight in gold. But the golden organ that seems like a source of wealth to a potential seller may well turn out to be a source of illth, for its removal can cause sickness and suffering in one who can least afford to be ill. In that case, the money received for the organ turns out not to be wealth, but illth.

As Ruskin has showed us, gold can bring us to the bottom of the sea. The commodification of human body parts smells of illth. When society gives approbation to a system in which desperately poor people begin to see their organs as commodities to exchange for other goods, something has gone terribly wrong with our vision of ourselves as human beings. Those desperate for money may not adequately consider their own health in deciding to sell an organ. The fact that selling organs is now illegal in the United States . . . reflects the definite sense of illth with which the selling of organs leaves us.

> "If xenotransplants were banned . . . the
> alternative is more human suffering and
> more deaths."

Animal Transplants May Solve the Organ Shortage

Steve Connor

The supply of organs for donation is vastly insufficient to meet the needs of recipients. Scientists have begun to consider using other animals, including pigs, as organ donors, in a process called xenotransplantation. although some people argue that xenotransplantation is cruel to animals, Steve Connor, in the following viewpoint, argues that the potential benefits of such transplants for humans far outweigh the harm done to the animals. Steve Connor is a contributing editor for the *Independent*, a British daily newsletter.

As you read, consider the following questions:
1. According to Connor, why did scientists choose pigs for xenotransplantation?
2. What does the author consider the "essence" of the problem with organ transplantation?
3. What beneficial breakthrough occurred in Cambridge, according to Connor?

From "Why Not Have a Heart-to-Heart with a Pig?" by Steve Connor, *The Independent*, August 1, 1998. Reprinted by permission of *The Independent* Syndication Department.

There is something deeply ingrained in the human psyche about the fear and revulsion inspired by the image of a creature that is half-man, half-beast. From the Minotaur to *A Midsummer Night's Dream*, the animal-human chimaera has been a grotesque spectacle relegated, thankfully, to the pages of mythology.

But could this always be the case? Now that the government has laid down the ground rules for the first organ transplants from animal to humans, it seems perfectly likely that the day will arrive when a man could be truly said to be a pig at heart. The first human patients to receive pig hearts could do so as early as 2000, though this optimistic timetable is just as likely to slip into the early years of the new millennium.

Many people will be understandably revolted by the idea of men or women walking around with pigs' hearts. The notion of the heart being the seat of emotional strength and fallibility goes back many centuries. The bravery of Richard I emanated from his lion heart, wicked witches throughout the ages were said to be heartless, and Cilla Black made her fame in the Sixties from asking whether anyone with a heart could not help but fall in love.

But this irrational fear of losing your heart to a pig, so to speak, cannot possibly justify the rejection of xenotransplantation, when animal organs and tissues are used in human patients. There is, nevertheless, a definite "yuk factor" associated with xenotransplants, especially when it comes to pigs. For many religions, the pig is considered an unclean animal which eats unspeakable things. But for anyone who knows pigs, there is something quite charming, if not unnerving, in the way they stare you right in the eye. As Winston Churchill once said, "dogs look up at you, cats look down on you, but pigs treat you as an equal."

Why a Pig?

Scientists have chosen the pig for very good reasons. It grows to about the same size as a human and so its organs can fit snugly into the space provided by a transplant surgeon. It is a domesticated animal which can be easily handled, and it is sufficiently different from humans to make the notion of using them more acceptable than, say, monkeys.

According to some experts, monkeys look a bit too much like us to be suitable xenotransplant candidates. In fact, they have already been used in the United States for a few unsuccessful transplant experiments on humans, but the research suffered from an image problem connected to the fact that monkeys look, and often behave, in a very human-like way. Higher primates, such as baboons and chimpanzees, share some of the characteristics—such as self awareness, fear and complex social organisation—that we often take to be uniquely human.

The poor old pig, however, does not get the same respect from the experts. "We realise that pigs are able to suffer but believe that their suffering is likely to be less than that of higher primates," explained Professor David Morton, a biological ethicist at Birmingham University who was a member of a committee investigating the issue, set up by the Nuffield Council on Bioethics. It seems the pig is just not self-aware enough to suffer pain in the same way as primates and humans.

It would, however, still be hard for people to raise ethical objections to using pigs if they are prepared to eat them. Carrying a pig-skin wallet next to your heart is, from the pig's point of view, just as bad as carrying a porcine heart inside your chest. If we are prepared to slaughter millions of pigs each year for their meat and skin, what could be the objection to killing (as humanely as possible) a few more to save thousands of human lives?

Rising Demand for Organs

This brings us to the essence of the problem. Only one in three people who need organ transplants receive them. As more people are living longer, the demand for replacement organs, as old ones wear out, continues to rise. Although there is continuing development of mechanical organs and artificial tissues and organs, this is unlikely to go anywhere near meeting the shortfall between supply and demand for at least another 20 or 30 years. In the meantime people, including children, are dying.

This means there is a desperate urgency to improve on

the previous attempts at xenotransplants. The most famous example is Fae, the "baby with a baboon heart" encapsulated in the lyrics of a song on *Graceland*, the Paul Simon album. But the days of miracle and wonder were not enough to save two-week-old Fae, who died within three weeks of receiving her primate heart. Her body's immune system had effectively eaten her new heart away in a violent chemical rejection. This type of rejection has continued to frustrate attempts at transplanting organs from one species to another.

The Era of Xenotransplantation

The era of xenotransplantation is here. . . . It is a 21st century answer to curing fatal diseases and prolonging life. Our society can no longer deny life-saving treatment to individuals simply because there are not enough human organs available. Xenotransplantation is a viable, exciting, and preferable option. The public will accept xenotransplantation better than the "presumed consent" system. Xenotransplantation may not fully replace human organs and tissue transplantation even with the recent successes of animal cloning. It can, however, serve as a back-up system for human graft donors. . . . We must be open to xenotransplantation as an alternative method that can eliminate the current human organ supply and demand problem.

Barbara Claxon-McKinney, *Pediatric Nursing*, September 2000.

Unlike normal organ transplants between people, xenotransplants pose unique difficulties associated with the nature of this violent rejection. Because the tissues of two species are so unrelated, the immune system does not merely reject them, it "goes nuclear" and attacks the invader with a powerful array of antibodies known as the complement reaction. For years, transplant scientists could do little to combat the complement reaction—until now.

Humanising Pigs

The breakthrough occurred in Cambridge where scientists were able to insert human genes involved in the complement reaction into pig embryos. The animals therefore grew up to have "humanised" organs that should, in theory, be less likely to be rejected. In 1995, the scientists showed that monkeys can survive with genetically-engineered pig organs for more than 60 days. This finally opened the doors to the use of pig organs in human transplant operations.

There was, however, one very real fear of rushing ahead. Many diseases, from AIDS to rabies, are known to involve the transfer of animal viruses to humans when the two have come into close contact. HIV is thought to have evolved from a similar virus of African monkeys. Pigs are also known to play an important role in the evolution of new and more deadly strains of the influenza virus. What could be the chances of an unknown pig virus getting into a human transplant patient and from there creating a new and lethal epidemic?

The answer, of course, is that nobody knows for sure. In 1997, scientists discovered a hitherto unknown retrovirus (the same viral group as HIV) in the pig which although it does not cause the pig any harm, can infect human cells in the laboratory, though this is not the same as causing disease in a person. It would be impossible to ensure a virus-free pig for transplants because a disturbing feature of these retroviruses is that they incorporate themselves into the animal's genetic material—its DNA—and so can therefore be inherited from sow to piglet, with scientists powerless to prevent it.

Playing God

Doomsters would argue that this is a good enough reason to stop everything now before we end up with a disaster. "This is potentially the first step down a very dangerous road indeed," said Roger Gale, a Conservative member of Parliament, on the announcement of the new government guidelines on xenotransplants. "Science really is now beginning to play God in a way that even those supportive of the advances of medical science must find disturbing, if not abhorrent."

Yet just because there is a risk attached to a new innovation, it is hardly grounds for junking the whole idea. Just about every medical advance, from vaccines to antibiotics, has carried risks, but the lives saved have made them worthwhile. The government guidelines address the concerns over animal welfare and establish the necessary framework for making sure—as much as anyone can—that the risks are kept to a minimum.

If xenotransplants were banned, as some animal welfare groups have proposed, then the alternative is more human suffering and more deaths. Animal rights activists should bear this in mind when they raise their objections. Their love of animals should not become a hatred of humanity.

> *"The potential for a worldwide disaster should be enough to stop any serious consideration of xenotransplants as a workable alternative to the shortage of human donor organs."*

Animal Transplants Will Not Solve the Organ Shortage

Alan H. Berger and Gil Lamont

To meet the growing demand for replacement organs, scientists have begun to genetically engineer donor pigs to make them more compatible with human recipients. In the following viewpoint, Alan H. Berger and Gil Lamont argue that xenotransplantation, or the transplanting of animal organs into humans, is unethical and poses too many dangers for society. They contend that deadly illnesses can manifest in human recipients of animal body parts and spread throughout society. Alan H. Berger is the executive director of the Animal Protection Institute (API), an organization that strives to protect animals from cruelty and exploitation. Gil Lamont is the managing editor of *Animal Issues*, the API's quarterly magazine.

As you read, consider the following questions:
1. Why do the authors claim that xenotransplantation will not be available to everyone?
2. What three diseases can humans acquire from animals, according to Berger and Lamont?
3. What incentives do the authors propose to increase the number of human organ donors?

Reprinted, with permission, from "Animal Organs Won't Solve the Transplant Shortage," by Alan H. Berger and Gil Lamont, *USA Today* magazine, November 1999. Copyright © 1999 by the Society for the Advancement of Education.

Imagine a world where anyone who needs a new heart, liver, or kidney can get it, a world in which there is no waiting list for an organ transplant or a problem with rejection. If this sounds too good to be true, it's because it is. Nevertheless, that is the picture the proponents of xenotransplants—the transplanting of animal organs into humans—want you to believe. They maintain that there is an unlimited supply of fresh organs, available to anyone in need. What they don't tell you is the downside of xenotransplants—the cost to the animals whose organs are used and to the humans who pay for it financially and ethically.

The Food and Drug Administration Xenotransplant Advisory Subcommittee has recommended the approval of "limited" human clinical trials for xenotransplantation. With the chronic shortage of human donor organs—the list of patients waiting for transplants is more than twice the number of organs available every year—the medical community has looked for a new source, and thinks it has found it in animals.

Imagine a world with a new kind of factory farming, where animal organs are "harvested" for hospitals and clinics, where genetically engineered pigs (now favored over non-human primates) only weeks old are taken away from their mothers and raised in antiseptic surroundings. Obviously, the medical community takes little interest in the animals' welfare. The sole value animals have is to be genetically altered to make them more "human," then used for spare parts. This appears to go completely against the medical community's own basic protocols of using "the fewest animals to benefit the most humans."

The Enormous Costs of Xenotransplantation

Despite what its supporters say, xenotransplants will not be available to everyone. Current transplant costs range from $116,000 for a kidney to more than $300,000 for a liver. Factoring in five years of follow-up charges puts the expense at nearly $400,000 for a liver transplant, with heart, heart-lung, and lung transplants running more than $300,000 each. According to the Institutes of Medicine (IOM), annual expenditures for transplants are about $2,900,000,000 annually. The IOM estimates that including animal-to-human transplants will push that figure to $20,300,000,000 a year, and it can only rise from there.

Billions of dollars that might have been directed to preventive medicine and education, to teaching healthy lifestyle practices such as diet and exercise and what to avoid—prevention that could save thousands of lives—instead will be devoted to the refinement and maintenance of xenotransplantation.

Prevention Is Key

Most illnesses are preventable. Studies show that changes in diet and increasing exercise can reduce high blood pressure, heart attacks, and cancer. Yet, many people behave as if their bodies are machines to be worn out, and the medical community sees transgenic pigs as the mobile warehouses for spare parts. Of the enormous numbers of dollars spent on health care, a fraction goes to prevention and control efforts—about three percent of most state public health departments' budgets. In 1994, more than $287,000,000 was spent at the state level on prevention efforts aimed at the six leading chronic ailments—heart disease, cancer, stroke, diabetes, obstructive pulmonary disease, and liver disease. This was approximately .07% of the estimated $425,000,000,000 spent annually to treat those same diseases, according to a Centers for Disease Control Study.

While transplants may offer longer and healthier lives to the chronically ill, "I really don't think transplantation is going to be the answer," indicates Charles Porter, a Missouri cardiologist. "It's going to be rehabilitation and prevention." Rep. Jim Moran (D.-Va.) agrees, pointing out that the nation spends far too much on curing illnesses and not enough trying to prevent them. "Prevention," he argues, "is much less expensive and far more effective."

Prevention has always been a better way, but most of the public doesn't take that view. As a consequence of xenotransplantation, health care premiums will rise; medical insurance coverage will become more limited; and the burden on Medicare will be astronomical. Fewer people will have health insurance and more will not be able to have access to adequate medical services. Currently, over 40,000,000

Americans are uninsured and around 34,000,000 face serious problems in obtaining needed medical care.

The Risk of Epidemic

The risk of transmitting infectious diseases to the wider population is a serious problem. We do accept the imposition of some risks on others for our own benefit (for example when we drive cars). However, there are limits to what we regard as acceptable (it is not acceptable, for example, to drive when drunk), and it is therefore necessary to consider the nature of the risk imposed on the wider population by xenotransplantation procedures. The difficulty for proponents of xenotransplantation is that the worst-case scenario (a major new epidemic) is extremely grave, and its likelihood is difficult if not impossible to quantify.

Jonathan Hughes, *Journal of Medical Ethics*, February 1998.

Imagine a world where taxpayers must pay crippling costs to support an already overburdened public health system and the massive expenses xenotransplants involve. Imagine a world where only the rich can afford transplants. Imagine a world where a previously unknown (and unrecognizable) virus is transmitted to a person through a transplanted animal organ, then races through the human population in a planetwide epidemic for which there is no cure.

Deadly Consequences

A deadly virus transmitted to humans from animals is not science fiction. It has already happened. The 1918 influenza epidemic that killed more than 20,000,000 people worldwide was a mutated virus of swine flu evolved from American pigs and spread around the globe by U.S. troops mobilized for World War I. Humans already can acquire approximately 25 diseases from pigs, including anthrax, influenza, scabies, rabies, leptospirosis (which produces liver and kidney damage), and erysipelas (a skin infection). Researchers recently discovered that human tissue cells are susceptible to infection by pig retroviruses. (HIV is a retrovirus, although it did not originate with pigs.) More than 30,000,000 people worldwide have been infected with HIV, a disease now accepted in research circles as having originated from animal to human contact.

These are not isolated examples. Health officials in Hong Kong destroyed millions of chickens after a virus, A(H5N 1), jumped directly from the birds to humans. There has been much publicity over the occurrence of Creutzfeldt-Jakob Disease, a fatal, brain-wasting illness that has claimed a number of British lives and has been linked to mad cow disease (bovine spongiform encephalopathy).

Even the U.S. government admits the danger. The 1996 U.S. Public Health Service xenotransplant draft guidelines include such alarming statements as:

- ". . . some agents, such as retroviruses and prions, may not produce clinically recognizable disease until many years after they enter the host. . . ."
- ". . . the full spectrum of infectious agents potentially transmitted via xenograft transplantation is not well-known."
- "Transmission of infections (HIV/AIDS, Creutzfeldt-Jakob Disease, rabies, hepatitis B and C, etc.) via transplanted human allografts has been well-documented."
- "As the HIV/AIDS pandemic demonstrates, persistent viral infections may result in person to person transmission for many years before clinical disease develops in the index case, thereby allowing an emerging infectious agent to become established."

In its 1996 report on xenotransplantation, the IOM tried to downplay the risk of an infectious disease being transmitted to the general public by assessing that risk as "greater than zero." How can anyone find this reassuring?

The potential for a worldwide disaster should be enough to stop any serious consideration of xenotransplants as a workable alternative to the shortage of human donor organs. Many researchers, doctors, and scientists across the nation have called for a moratorium on xenotransplants and human clinical trials. Yet, the government plunges ahead with xenotransplantation, a *Titanic* seeking an iceberg.

Motivated by Profit

What motivates this hasty and ill-conceived embrace of cross-species transplants? Maybe it is not the shortage of human donor organs so much as the potential for profit. One investment report estimates that, by 2010, just a decade away, 508,000 pig donors a year will be required worldwide. Annual revenues of pig organs will total more than $5,000,000,000. Pharmaceutical companies will earn at least that amount in annual rejection drug sales. With so much money at stake, it is not surprising that the biomedical community may be placing profits before public health.

Would xenografts even be an option if enough human donor organs were available? Recent advances point the way to greater compatibility of organ and recipient. The discoverer of a "facilitating cell" that helps stem cells engraft says that "universal organ transplants may be possible, which may alleviate the chronic shortage of organs." Meanwhile, there are some other potential solutions.

Possible Solutions

The Department of Health and Human Services should make tough policy recommendations on reallocating precious health resources. Health care services should be provided to patients who currently do not have access. This alone could save thousands of lives.

Serious incentives are needed to increase the number of human organ donors. These incentives can include education, marketing, changing the criteria for donor organs, and implementing a presumed consent law. The opposite of the present system, presumed consent would assume that everyone is an organ donor unless his or her refusal has been put on record.

The cry should be "Prevention! Prevention! Prevention!" Any doctor will tell you that preventive medicine dramatically reduces the risk of many serious diseases. Programs must be developed that encourage healthier diets, regular exercise programs, and stress management.

The National Institutes of Health needs to reallocate resource dollars marked for medical research. Too much money is spent on curing illnesses and not enough on trying to prevent them.

Xenotransplantation is not the answer, despite all the rosy pictures overoptimistic researchers, genetic engineers, and pharmaceutical companies paint. We cannot continue to cure human lives by the wholesale taking of animal lives. We cannot continue to deny health care to others simply because their economic state prevents health care from being "cost-effective." We must learn to take better care of each other, by becoming organ donors, and better care of ourselves, through diet and exercise.

> "Should we not use cells derived from
> donated embryos to save lives, just as we do
> after an auto accident by using the organs
> of those who tragically died?"

Fetal Stem Cell Research May Improve Medicine

Lawrence S.B. Goldstein

Stem cells have the ability to generate into any type of cell the body requires, and scientists claim that research into the use of stem cells may lead to improvements in the treatment of Alzheimer's disease, Parkinson's disease, and some types of cancer. However, because such cells are derived from aborted fetal tissue, ethical considerations have caused the federal government to halt funding for stem cell research. In the following viewpoint, Lawrence S.B. Goldstein argues that the potential medical benefits justify the use of stem cells; therefore, he advocates lifting the ban on government funding of stem cell research. Goldstein is a professor in the Division of Cellular and Molecular Medicine and Department of Pharmacology at the University of California, San Diego School of Medicine.

As you read, consider the following questions:
1. What are the problems with organ donation, as described by the author?
2. What differentiates human pluripotent stem cells from other cells in the body, according to Goldstein?
3. According to the author, what are three reasons why stem cell research should be supported by federal funds?

Reprinted, by permission of the author, from "Providing Hope Through Stem Cell Research," by Lawrence S.B. Goldstein, *San Diego Union-Tribune*, May 25, 1999.

Why should we use federal funds for human pluripotent stem cell research? Ask Walter Payton and 12,000 other Americans who are waiting for liver transplants. If they are fortunate, new livers will be found and they may live; if not, they will die.

Ask my friend Doug, who has a 7-year-old son with diabetes. Every night he and his wife are awake in the wee hours, monitoring their son's blood, worrying that they have missed the balance and that their beloved child will slip into a coma.

Ask the children of millions more like them, for whom insulin is a treatment but not a cure, because crucial cells in the pancreas are still missing. These children are always in danger, and they live under the constant shadow of premature death or disability.

New hope for these desperately ill people has come from the recent discovery of "human pluripotent stem cells," the primordial cells from which all the tissues and organs of the body develop. However, a serious debate has recently erupted on Capitol Hill about whether federal funds should be used to support further research in this area. At issue is whether the merits of public funding and the dreadful burden of disease balance concerns about the origin of these special cells.

The Limitations of Organ Donation

To understand the need for research with human pluripotent stem cells, one need look no further than many common diseases such as cancer, heart disease and kidney disease. These diseases are treatable in whole or in part by tissue or organ transplants, but there are persistent and deadly problems of rejection and a completely inadequate supply of suitable donor organs and tissues.

In addition, the grim arithmetic of most organ transplants is that those who are seriously ill wait for the tragic accidental death of another person so that they may live. Worse, for juvenile diabetes and many other diseases, there is not even a suitable transplantation therapy or other cure.

The recent discovery of human pluripotent stem cells has suddenly given us the potential to escape these dilemmas and provide our most desperately ill children, friends, parents and neighbors with new tissues and organs to replace their own damaged ones. Pluripotent human stem cells, unlike all other cells in the human body, seem to "remember" how to become almost any type of cell or organ.

Our scientists are at the threshold of learning how to coax these cells into growing into the many kinds of organs and tissues needed by our gravely ill citizens, without the potential problems of rejection seen in most transplants. Thus, our researchers may soon be able to generate pancreatic cells to save my friend Doug's son, and liver cells to rescue Walter Payton and those like him.

Possible treatments for Alzheimer's disease and many cancers may also be forthcoming. Federal funding for research with pluripotent human stem cells is desperately needed for our scientists and physicians to realize these worthy goals.

Ethical Concerns

Although human pluripotent stem cell research has tremendous medical potential, some of our citizens and legislators seek to prohibit our best and brightest federally funded university scientists and physicians from working with human pluripotent stem cells. They do so because of ethical concerns about the origins of these cells, which were derived from the earliest human blastocyst stage embryos.

While I, and many other scientists, share these concerns, we believe that the potential benefit of such work and other ethical considerations balance the concerns about the origins of these cells. In fact, 33 Nobel laureates recently wrote to the president and the Congress in support of federal funding for research with human pluripotent stem cells to ensure that those with deadly diseases are given this chance.

There are several major reasons to believe that federal support of human pluripotent stem cell research is both appropriate and ethical:

- Using federal funds to support pluripotent stem cell research guarantees proper ethical oversight and public input into this important work. Federal funding of this research will require the scientific community and the government to work together to establish an appropri-

ate set of rules for this research. These rules will ensure the advancement of critical medical research and maintain respect for public sensibilities.

The National Institutes of Health has already developed an outline of such a system. For federally funded research, this system will prohibit the use of pluripotent stem cells that have come from "embryo farms" or from embryos purchased or sold, and will continue to ensure that human embryos are not created for research purposes. It will also continue to be illegal to use federal funds to derive pluripotent stem cells from human embryos.

- Banning federal funding for human pluripotent stem cell research will not eliminate it. Such research will proceed in private industry and in other countries. This fact prompts serious concern that the work may then be conducted in secret, without the benefit of ethical regulation or public debate as it proceeds.

- Using federal funds for human pluripotent stem cell research ensures that our best and most capable scientists will participate in this research. Without such funding, new treatments will be delayed by years, and many who might otherwise have been saved will surely die or endure needless suffering.

- Federal funding is the best way to guarantee that stem cell therapies are developed with the greatest consideration of the public good. In the absence of federal funding, it is likely that stem-cell derived treatments will only be found for diseases that commercial companies determine will yield the largest profit if treated. Thus, market forces could create a situation where deadly, but less profitable, diseases are ignored.
- Although it is essential that we use federal funds to support pluripotent stem cell research, the stem cells themselves will be derived without using federal funds from early embryos that are destined to be discarded. In vitro fertilization treatments for childless couples often produce more embryos than can be implanted into the mother. These embryos cannot develop on their own, have only a few cells, and must either be stored in freezers indefinitely, or eventually destroyed.

  There is no other ethical use for these embryos if the parents choose not to have them implanted into the mother. Even if one believes that the destruction of these embryos is a tragedy, should we not allow the parents the right to make the decision to donate them for pluripotent stem cell derivation and stem cell research so that many other people might live? Should we not use cells derived from donated embryos to save lives, just as we do after an auto accident by using the organs of those who tragically died?
- Balancing the ethical objections of some to pluripotent stem cell research are the serious ethical implications of not proceeding. Can we justify turning our backs on our children, parents and friends who will suffer and die if we do not find suitable cures? Ethically validated pluripotent stem cell research provides new hope for these people.

In the past few years, Congress has wisely and dramatically increased federal funding for biomedical research. We must ensure that these funds are used for the best and most promising medical research.

Ethics, scientific opportunity and medical need can surely be balanced. Ask Walter Payton or my friend Doug if you're still not sure.

"*We're progressing to a civilized new stage—turning human beings into valuable commodities—in which the bodies of the helpless are used to improve the lives of the powerful.*"

Fetal Stem Cell Research Is Unethical

John Kass

Stem cells, or human pluripotent cells, have the ability to develop into any kind of cell in the body. Much controversy has arisen regarding the ethical implications of using these cells—which are harvested from aborted fetuses—to treat medical conditions. In the following viewpoint, *Chicago Tribune* columnist John Kass argues that the use and research of fetal stem cells is unacceptable because it requires exploiting human beings (fetuses) who have not consented to being used in this manner.

As you read, consider the following questions:

1. How does someone with Parkinson's disease resemble a fetus, in the author's opinion?
2. According to the author, what effect did the new brain stem cells have on Jacqueline Winterkorn's brain?
3. Why does the author claim that aborted babies will become resources?

It's an ugly twist on an old science fiction theme:

Would you use the body parts of an innocent baby so that you could live a happier life?

Would you support a system of incentives to kill other babies, and process them like meat at a packing plant, for the benefit of a frightened Baby Boom generation terrified of Alzheimer's disease and death?

Of course not. The suggestion is monstrous and dehumanizing. By comparison, it makes what the Serbs and Albanians are doing to each other look like a gentle game.

But the science fiction scenario doesn't generate the terrifying passions of old Balkan blood feuds.

Instead, it's calculated, without anger, and practiced by reasonable men and women in white lab coats.

It's about pure reason, efficiency and scientific rationalism. It's what a culture can do when it loses its soul. If you don't believe me, ask a Jew about the Nazi concentration camps.

A Science Fiction Nightmare

So get horrified. Because it's not science fiction. It's happening now, in our country.

I read about it in the *Chicago Tribune* on June 27, 1999, in a fascinating story by science writer Ronald Kotulak under the headline "Stem cells opening path to brain repair."

It began with an anecdote about a woman with Parkinson's disease. Her name is Dr. Jacqueline Winterkorn. The drugs she was taking to fight the disease weren't working anymore.

"It's a very sad disease," Dr. Winterkorn was quoted as saying. "People are locked into bodies that don't move. Their brains are working, their minds are working, but they can't talk and they can't move."

In other words, they're human beings immobilized through no fault of their own, trapped without speech. They have emotions, but they can't do anything about it. They're helpless.

Like a fetus.

But Dr. Winterkorn's condition began improving, the story said, after she was given millions of new brain stem cells because her own brain cells weren't doing their jobs. Her brain cells weren't producing enough dopamine to control her movements.

The new brain stem cells worked just fine. They produced dopamine in her brain. She improved. The scientists are thrilled.

"The prospect of repairing a damaged brain is pretty remarkable," said Dr. Curt Freed, who did the study. "It has been possible to show significant improvements in some patients who suffered from a chronic neurologic disease for an average of 14 years."

But there is a price for Dr. Freed's success. The new brain cells have to come from somewhere. And they don't come from pigs.

Aborted Fetal Tissue

They come from fetuses, which is a polite way of saying they come from tiny human beings. The tiny human beings didn't willingly give up their brains. Nobody asked them to sign papers donating their bodies to science.

They didn't have much say in the matter. They were aborted.

The National Institutes of Health—which means the federal government—has lifted its ban on the use of human fetal cells and is bankrolling several other similar studies.

Meanwhile, the White House worries that video games

Embryonic Research Is Immoral

Even from a scientific or ethical perspective human embryo experiments are unacceptable. The basic science that is used to determine the "moral status" of these early human embryos is grossly incorrect. There is absolutely no question whatsoever, scientifically, objectively, that the life of every human being begins at fertilization. There is no question philosophically that any attempt to split a human being from a human person is both theoretically and practically indefensible. Personhood begins when the human being begins—at fertilization.

Therefore, any experiment which would require the intentional destruction of innocent human beings—even if for the greater good of society, or for the advancement of scientific knowledge, or the national security—is automatically unethical. Great benefits do not justify unethical means.

Dianne N. Irving, "NIH and Human Embryo Research Revisited: What's Wrong with This Picture?" 1999.

---

cheapen human life and make possible massacres like the one in Littleton, Colo.

Courts and abortion rights advocates have said that what grows in a mother's womb is not a human being. You don't say baby. That's impolite. You say "it," because that makes a human being easier to kill.

The debate over abortion is an old one now. Most folks have settled into their positions and defend them vigorously. That's not going to change.

What's changing is that we're progressing to a civilized new stage—turning human beings into valuable commodities—in which the bodies of the helpless are used to improve the lives of the powerful.

And it's being done in the name of cold scientific reason. The rhetorical pathway was cleared years ago, when the Germans built Buchenwald and Auschwitz and other places.

Soon other folks with Parkinson's or other brain disorders such as Alzheimer's disease will seek such treatments. The Baby Boom generation that has never been denied will make its demands.

The Survival Instinct

It's human nature to use available resources to satisfy the most powerful human need: staying alive.

So aborted human babies will become resources. They'll become products, subjected to the market. Because they'll have value, there will be an incentive to provide more. Their bodies will be served up for the benefit of adults.

If we don't stop it now, if we accept this crime in the name of scientific reason, we'll lose ourselves.

Ask a mother carrying a child inside her. Ask her if it's not human. Ask any father who puts his hand on his expectant wife's belly and feels a tiny foot.

In a few weeks, they're out and looking up to you. They grab your finger. You kiss their necks. Someday, when they're old enough, they might ask you what fetal brain stem cell research is all about.

What will you tell them?

Periodical Bibliography

The following articles have been selected to supplement the diverse views presented in this chapter. Addresses are provided for periodicals not indexed in the *Readers' Guide to Periodical Literature*, the *Alternative Press Index*, the *Social Sciences Index*, or the *Index to Legal Periodicals and Books*.

Ron Brown	"A Free Market in Human Organs," *Freedom Daily*, February 1996.
Barbara Claxon-McKinney	"Xenograft Technology: A New Frontier in Medicine," *Pediatric Nursing*, September 2000.
Judith A. DePalma and Ricard Townsend	"Ethical Issues in Organ Donation and Transplantation: Are We Helping a Few at the Expense of Many?" *Critical Care Nursing Quarterly*, May 1, 1996.
John J. Fung	"Transplanting Animal Organs into Humans Is Feasible," *USA Today*, November 1999.
Diane M. Gianelli	"Stem Cell Research Focus of Ethical Dilemma," *American Medical News*, December 21, 1998.
Jonathan Hughes	"Xenografting: Ethical Issues," *Journal of Medical Ethics*, February 1998.
Paul Likoudis	"Dead Baby Parts Business Booming," *Wanderer*, September 30, 1999.
Jonathan D. Moreno	"The Dilemmas of Experimenting on People," *Technology Review*, July 1997.
Nancy Scheper-Hughes	"Postmodern Cannibalism," *Whole Earth*, Summer 2000.
Harold Y. Vanderpool	"Xenotransplantation: Progress and Promise," *Western Journal of Medicine*, November 1999.
Robert J. White	"Human Embryo Research," *America*, September 14, 1996.
Susan Wills	"A New Growth Industry in Baby Body Parts," *National Right to Life News*, November 1999.

CHAPTER 3

How Has Technology Affected Privacy?

Chapter Preface

Technological advances such as hidden cameras and monitoring software have enabled employers to observe employees' activities in the workplace. Courts have recently found in favor of employers sued for reading e-mail, listening to voice-mail messages, and tracking Internet usage. While many contend that employee monitoring allows employers to maintain productivity, others argue that such measures violate employees' rights to privacy.

Privacy rights advocates argue that some employers fail to fairly notify employees of monitoring policies. According to the American Civil Liberties Union (ACLU), "An employer may tap an employee's phone line, may watch his employees though a secret camera, may read his employee's electronic mail, may search though his employee's computer files, all of this without the employee's consent. The employee does not even have to be told that [he is] being monitored." Supporters of employees' rights to privacy contend that such monitoring is not only disrespectful toward employees, but also aggravates tension between employees and management.

Employers argue that not only are valuable company time and resources eaten away by employees' surfing the Web or chatting on the phone, but the company may also be liable for the distribution of offensive material such as hate mail or pornography around the workplace. "It raises all sorts of concerns about sexual harassment, racism and other issues. Companies face problems with people sending an off-color joke, and if the company knows or should know about it, it's going to be held liable for letting that kind of conduct go on," according to editor Gillian Flynn. Employers maintain that monitoring employees protects the interest of the company.

Employee monitoring in the workplace is one of the issues discussed in the following chapter on how technology has affected privacy.

| *"Technology has changed the rules of privacy."*

The Internet Threatens Privacy

Jeffrey Rothfeder

In the following viewpoint, Jeffrey Rothfeder argues that the increasing convenience of technology and computers has come with a price: the loss of personal privacy. The Internet, which many regard as anonymous, is riddled with devices that track which websites users visit. Marketers then use that information to target prospective clients or sell it to other companies for a profit. Rothfeder claims that users must be aware of their lack of personal privacy on the Internet and guard against unwanted intrusion or potential fraud. Rothfeder is the author of four books, including *The People vs. Big Tobacco* and *Privacy for Sale*.

As you read, consider the following questions:
1. According to the author, how have technology and personal computers changed the rules of privacy?
2. What three types of computer files are protected by law, according to Rothfeder?
3. As cited by the author, what were the recommendations of the Commerce Department in response to the 1998 Federal Trade Commission report?

Excerpted from "You Are for Sale," by Jeffrey Rothfeder, *PC World*, September 1, 1998. Reprinted with permission.

M edical histories, bank balances, even unlisted phone numbers—the details of your life are brokered online every day.

A Washington, D.C., man who is desperately trying to quit smoking receives a letter from a marketing firm: "Our records indicate that you have tried to stop smoking using a prescription nicotine replacement product. We hope you have succeeded, but if you, like many others who have tried to quit, are still smoking, we have good news for you." The rest of the letter touts a new cessation drug called Zyban.

Meanwhile, a week before her birthday, a Pennsylvania woman gets a card in the mail from RadioShack, wishing her happy returns and offering $10 off her next purchase at one of the company's retail stores. She has never bought anything from the store and never told RadioShack when she was born. No matter. RadioShack bought the information from the state's motor vehicles department.

Technology has changed the rules of privacy. We go through life inadvertently dropping crumbs of data about ourselves. Following right behind us are powerful vacuum cleaners—computers accessed by marketers, snoops, and even criminals—sucking up the crumbs, labeling them, and storing them for future reference.

"Privacy is like clean air," says Kevin Murray, who runs Murray Associates, a New Jersey–based firm that sweeps clients' offices for bugs and other surveillance equipment. "At one time there was plenty of it. Now it's almost gone."

A Long Time Coming

Long before the Internet and home PCs became staples of everyday life, credit bureaus and junk mailers collected information about consumers from the purchases they made and the warranty cards they mailed. But gathering this data was slow and expensive. Most of it had to be input by hand, making the task of cross-referencing other information sources virtually impossible.

All that has changed. Personal computers, sophisticated database software, and electronic information networks have transformed the slothful business of poring over mainframe records into a high-tech industry that compiles, cross-references, and exchanges private data instantaneously.

At the same time, the amount of available information about us has increased astronomically. When we buy books and groceries, rent movies, or pay bills with credit and debit cards, we give away information about ourselves. Each time we visit a Web site, we unwittingly leave traces of who we are. As a result, data gathering (or data mining) has become a booming business, with scores of firms pooling what they know about us.

Wells Fargo Bank, for example, has teamed up with a grocery chain in California (the bank declines to say which one) to cross-reference people's shopping patterns with their financial records—a process called cluster analysis. This helps the bank predict which of its services a person may be interested in, based on shopping habits.

"The aim [of the project] is to be able to promote, for instance, a self-directed IRA to everyone who makes over $50,000 and buys Arabica bean coffee every week," says a Wells Fargo Bank marketing vice president, who requested anonymity.

The top data aggregators—companies like Metromail, First Data Solutions, and Acxiom—each maintain information on more than 90 million households and 140 million people. Their databases store such tidbits as when we were born, how often we travel, what we buy, which prescription drugs we use, and whom we call.

"In a perfect world, companies would have to get people's consent before they shared information about them," says Evan Hendricks, editor of the Washington, D.C.–based newsletter *Privacy Times*. "But this isn't a perfect world."

The most obvious results of data mining are extra junk mail and unrelenting telemarketers who call at dinnertime. But this free flow of information can also have sinister consequences. The things you tell your doctor could keep you from getting a job. Your reputation can be made or lost depending on what your electronic profile—accurate or not—says about you. And an unlucky few may have their identities stolen by computer-age criminals who obtain victims' credit files and make purchases in their names, leaving them in financial ruin.

Money Talks

Among the biggest information databases are those maintained by the three major credit bureaus—Experian (formerly TRW), Equifax, and Trans Union. They contain a wealth of information about people's income, jobs, bank accounts, purchasing behavior, and credit limits. An aggressive company such as Experian can combine this data with public information that it draws from motor vehicle and property records. It can then put together sophisticated lists with the names of individuals who, for instance, live in the Dallas area, make more than $100,000 per year, drive foreign cars, and have no more than two children.

But the greatest danger is what can happen when credit data falls into the wrong hands.

Credit reports are one of only three types of files protected by federal law (the other two are video store records and cable television accounts). Only you and any third parties you've authorized (such as your landlord) can access your report. But with so many resellers marketing credit files online, a data criminal can easily obtain unauthorized files, open new credit-card accounts, get loans in your name, or attempt to blackmail you.

"If you walked into a bank and wanted a $2000 loan, [the bank] would want every bit of information about you, including your shoe size," says David Szwak, an attorney based in Shreveport, Louisiana, who specializes in credit-card fraud and computer security. "But if you know enough about someone, such as his social security number, electronically you can get a $5000- or $10,000-limit credit card in his name with no personal contact."

That's what happened to Ken Robinson, a client of Szwak's near Dallas. Someone stole Robinson's identity and bought a satellite dish, a diamond ring, and a houseful of furniture in his name. Two years later, Robinson is still trying to clear his record. "It's just a never-ending nightmare," he says.

How easy is it to get a credit report? Fill in the requisite information at QSpace, and within minutes you can get a person's Experian credit file on-screen. QSpace, a company based in Oakland, California, claims that by using VeriSign's encryption technology it can ensure that only authorized persons may view an individual's credit report.

But accidents do happen. In 1997, Experian began offering credit reports via e-mail from its Web site. The company discontinued the program after just one day when a glitch in the system sent credit reports to the wrong e-mail addresses. At the time, Experian said it would restart its online credit report program as soon as it had worked out the security bugs; that day hasn't come yet.

Back then, the Experian incident scared away Equifax and Trans Union from providing online reports. But recently, Equifax launched an initiative with IBM to use an electronic certificate system to make sure that online credit reports go only to authorized recipients.

QSpace won't discuss its security procedures. "Due to the competitive nature of the Internet, spelling out details of our security policy would not be in our best interest," says Arash Saffarnia, QSpace's chief technology officer.

Unhealthy Disclosure

As sensitive as credit reports are, medical records are even more so. But here, too, your privacy is at risk. The biggest repository of medical files in the United States and Canada is the Medical Information Bureau (MIB). This consortium of insurance companies maintains millions of records culled from insurance applications as well as from doctors' and hospitals' files. When someone applies for a policy, insurers scan MIB's computers for information about any preexisting conditions that might affect their decision to issue the policy or how much to charge.

MIB tries to be careful about who sees its files. But because every insurer in the United States and Canada has access to these records, information sometimes ends up in the wrong hands. Paul Billings, chief medical officer for a Texas health care network, wrote about such a case. He tells of a woman who was turned down for a job because her MIB file indicated that she had a predisposition for a muscular disease, even though she had no symptoms of the condition. Her prospective employer had obtained her records through the firm's insurance company.

Lately, marketers have gotten into the act, increasingly using medical records to target consumers. For example, Elensys, a company in Woburn, Massachusetts, manages electronic records for thousands of pharmacies and uses their customer files for marketing purposes. Developing mailing lists on the basis of prescription information, Elensys sends personalized letters—on pharmacy letterhead and sometimes paid for by drug manufacturers—reminding customers to keep taking their medicine or pitching new products to treat an ailment.

Opening Up to Strangers

The sociologist Georg Simmel observed nearly 100 years ago that people are often more comfortable confiding in strangers than in friends, colleagues or neighbors. Confessions to strangers are cost-free because strangers move on; you never expect to see them again, so you are not inhibited by embarrassment or shame. In many ways the Internet is a technological manifestation of the phenomenon of the stranger. There's no reason to fear the disclosure of intimate information to faceless Web sites as long as those Web sites have no motive or ability to collate the data into a personally identifiable profile that could be disclosed to anyone you actually know. By contrast, the prospect that your real identity might be linked to permanent databases of your online—and off-line—behavior is chilling, because the databases could be bought, subpoenaed or traded by employers, insurance companies, ex-spouses and others who have the ability to affect your life in profound ways.

Jeffrey Rosen, *New York Times*, April 30, 2000.

Elensys spokesperson Kathryn St. John defends the company's practice as a service to patients. She adds, "It's important to note that we don't sell the names in the database to other companies so they can market products."

But some critics aren't appeased. "It's a gross invasion," complains George Lundberg, a physician and the editor in chief of the *Journal of the American Medical Association*. "Do you want the great computer in the sky to have a computer list of every drug you take, from which can be deduced your likely diseases—and all without your permission?"

In the shadow of big credit bureaus and medical consortiums are dozens of smaller online firms that sell all types of personal data. Some of these sites, like www.whowhere.com and www.switchboard.com, are relatively innocuous repositories of names, addresses, and phone numbers. The more disturbing sites deal in more invasive forms of personal information.

Companies like 1-800U.S.Search, American DataLink, A1-Trace USA, Discreet Data Systems, and Dig Dirt trumpet their wares on Web sites. Enter a social security number at 1-800U.S.Search, and within an hour you can get someone's current and past addresses for up to ten years, as well as telephone numbers, date of birth, and aliases. And if you want a background report on "nannies, employees, associates, doctors, neighbors, or friends," the company will provide, among other things, driver's license information, vehicle ownership, and bankruptcies. Go to A1-Trace's site and you can dig deeper. For $179, find out what's in someone's safe deposit box; for $289, access bank records; and for $789, learn how much a person has saved in overseas accounts.

These data resellers are usually run by private investigators, former cops, or ex-corporate security chiefs. Legitimate companies are circumspect about what they provide customers. They don't give out credit reports without authorization, for instance, and they may withhold social security numbers if they question a buyer's motives.

And what they sell isn't illegal. No federal law will safeguard your medical files, bank accounts, phone logs, or phone numbers, so such data can be sold without fear of prosecution. In fact, most federal and state agencies sell motor vehicle records, voter registration files, and other data to information resellers.

Dangerous Characters

There is, however, an underground network of resellers who freely traffic sensitive data, no matter how personal or illegally obtained. These firms collect information from dozens of legal as well as illicit institutional sources (including the major credit bureaus), in addition to informants at banks, insurance companies, and other firms.

In 1992, the federal government cracked down on such

criminal activity. But the perpetrators—Super Bureau, Tracers Worldwide, and six other companies—got off with the equivalent of a slap on the wrist: in most cases, a one-year suspension and a warning. According to some legitimate information resellers, who requested that their names be withheld, most of those companies are currently back in business on the Internet.

"Being able to peddle their wares on the Internet lets them sell information quickly, hidden behind an electronic screen, and move on to the next job before any curious authorities look too closely," says an insider.

All of this secretive activity on the Internet has the authorities stymied. "So many data companies come and go, change their names, close their sites, and open new ones on the Internet," says one FBI agent who heads up a computer crime unit. "Only if they do something egregious and obviously illegal, like selling tax returns, can we hope to stop them."

Net Surfers Anonymous?

But as the Internet grows, the worst threat to privacy may ultimately be consumers' own lack of discretion. Each day hundreds of people post messages to Usenet newsgroups, offering personal information as if they were talking to a friend in their living room. In reality, each piece of electronic correspondence can be seen by millions of Internet users for many years to come.

"Usenet droppings," as electronic-information resellers like to call them, are among the most fertile sources of personal information for digital sleuths. Karen Coyle, western regional director of Computer Professionals for Social Responsibility in Palo Alto, California, recalls an incident in which a woman complained in a health care newsgroup about her medical plan and physician. Soon thereafter, she received an e-mail from a representative of her health maintenance organization, asking her if she would like to discuss why she was unhappy with her plan. It seems that her HMO scans newsgroups for all mentions of itself, and then follows up by responding to comments.

None of this sits well with Americans, who are becoming increasingly frightened about their lack of online privacy. A

Business Week/Harris poll conducted in 1997 found that 53 percent of the respondents believe that laws should be passed to specify how personal information can be collected and used on the Internet.

So what's being done to combat this invasion of privacy? Because of the relative novelty of Web-based data collection, the Internet has been the target of most initiatives. In June 1998, the Federal Trade Commission issued a report claiming that Web sites are not doing enough to protect surfers' privacy.

Soon after the FTC's report came out, the Commerce Department proposed guidelines for safeguarding people's private information online. Among the Commerce Department's recommendations: Web sites should disclose when they are collecting information about users and outline what they will do with the data; visitors should also be given the option to determine how the information may be used; medical records should not be shared without a patient's consent; and companies should be held accountable when privacy policies are violated. . . .

However, nothing will keep snoops and data miners from making money off your personal information in the foreseeable future. Therefore, the burden is squarely on you. Whether you're online or offline, learn how your personal information can fall into the wrong hands and what you can do about it. . . . Maybe you won't mind getting a birthday card from RadioShack. But don't wait until RadioShack sends you a mysterious bill for $500 before you decide to take action.

> "*Our 'current concept' [of privacy] is evolving with cultural trends that help render some privacy questions irrelevant.*"

Technology May Not Threaten Privacy

Travis Charbeneau

According to freelance writer and futurist Travis Charbeneau, technological advances in information and genetic technology are raising new privacy concerns over such issues as health and financial information. At the same time, he contends, society is becoming less judgmental about issues that were once considered extremely private, such as homosexuality, making privacy less essential. Charbeneau argues that institutional reform can eliminate the need for privacy on many issues. These changes, along with software, can adequately protect privacy.

As you read, consider the following questions:
1. What is "Carnivore," as cited by the author?
2. According to the author, what are three previously private issues that have become more publicly acceptable?
3. What does Charbeneau consider the most effective pathway toward guaranteed privacy?

Reprinted, with permission, from "The Future of Privacy: Moot?" by Travis Charbeneau, *Mindspring.com*, January 5, 2001, available at www.richmonder.com/charbeneau.

In 2000 the federal government introduced us to "Carnivore," a not-so-reassuringly-named software utility that dragnets our email streams looking for fish to fry. As information technologies generally become more invasive and the information they collect harder to control, we justifiably fear for our privacy. Typically on the sly, in what one popular new book calls "The Unwanted Gaze," government, corporations and other entities increasingly track our physical activities and rummage through what have become virtual extensions of our brains: our computers and Internet presence.

Combining this data with what others collect, these agencies may eventually gain access to everything from the contents of our urine to our credit histories. Such privacy violations are bad enough when the resulting files are accurate and used responsibly, and, of course, they're often wildly wrong and criminally abused.

Worse, once digitized, this data is more difficult to control than common gossip. "The right to privacy," as deduced by Earl Warren's Supreme Court from the "penumbra" of the Constitution, has never faced such an assault, with volatile technologies fast outstripping our ability to regulate or even to understand them. It becomes questionable whether our current concept of privacy can survive this onslaught.

Happily, our "current concept" is evolving with cultural trends that help render some privacy questions irrelevant. Consider homosexuals "coming out of the closet," a trend that has certainly tended to render "exposure" obsolete. Secrecy here used to be critical for employment, avoiding blackmail, or even staying out of jail.

Exposure Has Become Commonplace

To the extent that we've moved away from poisonous attitudes respecting sexual orientation, "exposure" has become irrelevant. No, we aren't there yet, but the day is coming when sexual orientation will be considered as matter-of-fact as hair color. Bleached blond jokes aside, we normally aren't concerned about keeping our natural hair color a secret.

Likewise, in the recent past so much as a whisper of mental illness anywhere in the family, let alone evidence of any personal affliction, were matters of dire concern. Exposure of treatment for problems as common as mild depression could mean the end of a career—or never getting a job in the first place. But, as with homosexuality, cultural attitudes are fast in the process of shifting away from irrational condemnation.

Computers Make Crimes Easier

Cyberspace offenders—ranging in age from preteen to senior citizen—will have ample opportunities to violate citizens' rights for fun and profit, and stopping them will require much more effort. Currently, we have only primitive knowledge about these lawbreakers: Typically, they are seen only as nuisances or even admired as innovators or computer whizzes. But increasingly, the benign "hacker" is being replaced by the menacing "cracker"—an individual or member of a group intent on using cyberspace for illegal profit or terrorism.

Access to cyberspace has begun to expand geometrically, and technology is making the information superhighway even more friendly and affordable for millions of users. But foolproof protective systems can probably never be developed, although some high-tech entrepreneurs are certainly trying. Even if a totally secure system could ever be developed, it would likely disrupt the free flow of information—an unacceptable intrusion to most users. In fact, it is the ease of access that is driving this rapidly expanding field of crime.

Thomas E. Weber, *Wall Street Journal*, June 27, 1996.

Respecting substance abuse, those who have been through The Betty Ford Center have far less reason to hide than the addicts of 20 years ago. Same with divorce, adultery, illegitimacy, bankruptcy, and interest in pedestrian pornography. To some, increasing forbearance for such human foibles heralds the End of Western Civilization. To most, however, growing tolerance demonstrates growing understanding of the admonition against casting stones and a wiser appreciation generally of "There but for the grace of God. . . ."

To a very great extent, therefore, the quality of society determines its nature and need for privacy. In a society wracked by self-righteousness, ignorance and fear, privacy generally is critical. The more informed and enlightened the realm, the fewer the secrets that need keeping. Naturally and ideally, only genuine character issues and overt behavior should remain at issue. Arguably, if one has been a persistent liar or child molester, exposure remains appropriately hurtful.

The old bromide goes, "Never do or say anything you wouldn't mind seeing in the next day's newspapers." Few of us live lives quite so spotless. On the other hand, even an interesting variety of spots, again, short of genuine ethical offenses and outright crimes, shouldn't count for much in a society of more enlightened newspaper readers. Titillation may be with us forever, and to be its subject forever annoying. But condemnation is the crucial issue.

Protecting Traditional Privacy Rights

The essential question respecting the current assault on privacy is, can trends for greater tolerance, combined with vigilant legislative constraints protecting traditional privacy rights, outstrip technology's fast-growing power to invade and harm? Relatedly, we are also questioning the structure of some institutions.

With DNA profiling, for example, we gain a growing ability to predict who will get what diseases. But this presents a real privacy danger only in America, where the first order of our health care system is not to care for health, but to eliminate profit risks. Do we sweat bullets trying to keep insurers from finding out who is and is not a risk? Or do we follow the rest of the industrialized world in obtaining a system that unconditionally protects everyone?

Do Swedes predisposed to arthritis worry that some prospective employer will find out about it? In this case the enlightenment required applies not to any irrational social stigma respecting health, but to irrational social contracts. The need for privacy can be obviated in some cases by institutional reform.

Elsewhere, government needs to act quickly and decisively to keep privacy legislation and court decisions up to speed. At the user end, wonderfully secure encryption software and anonymity procedures are already available for email, newsgroup postings, Web browsing. These protections need to be made more widespread and user-friendly. Perhaps foremost, however, continued progress towards a more human-friendly society will do more than anything else to guarantee privacy where it's really needed.

A simultaneously more open and open-minded society enables us to shrink our respective privacy spheres. A smaller, more manageable privacy sphere, safeguarding only those issues that remain genuinely sensitive, means more certain protection irrespective of technological advance.

"Companies . . . no longer hesitate to seek information on what was once assumed to be the private side of workers' lives."

Workplace Monitoring Violates Employee Privacy

Dana Hawkins

Technological advances have made it easier for employers to monitor their employees' activities both in and outside the workplace. Installing hidden cameras, reading e-mail, tracking Internet usage, and genetic testing all help the employer keep tabs on workers. In the following viewpoint, Dana Hawkins argues that this surveillance violates employees' rights to privacy. She contends that employees' privacy rights ought to be respected and observed by employers. Hawkins is a senior editor at *U.S. News & World Report*.

As you read, consider the following questions:
1. According to Hawkins, what is the purpose of a "midnight raid"?
2. As described by the author, what is a major concern surrounding drug testing?
3. Why is employee privacy not protected by the Fourth Amendment, as described by Hawkins?

In the secrecy of night, a pair of private detectives meet their client at the back entrance to the communications company he owns. The man opens the door; they all slip inside and proceed to a worker's cubicle. The investigators sift through his file cabinets, desk calendar, Rolodex, and voice mail. They check a company caller-ID machine that recorded all the numbers he called in the past six weeks. Then they "bag" his computer, downloading information from the hard drive, where the most damning data usually reside. One of the detectives is a forensic computer analyst, who will recover E-mails and documents long ago erased. With any luck, they will find what they are looking for: evidence that the worker gave proprietary information to a competitor.

The incident described above—known in the gumshoe trade as a "midnight raid"—is true. Similar intrigue takes place regularly in workplaces across the country, for a variety of reasons: to collect evidence to fire an employee, to defend against a discrimination lawsuit, to catch a company thief. "We do them all the time," says Christopher Marquet, senior director for global development at Kroll Associates, a leading investigative firm. "We have 22 offices worldwide, so there's probably always one going on somewhere." (Kroll was not the agency involved in the case above.) Employee investigations—which sometimes lead to a raid—have increased an average of 30 percent each year since 1994, according to Kroll and its competitors.

Keeping Watch

Companies not only have stepped up midnight raids but no longer hesitate to seek information on what was once assumed to be the private side of workers' lives. More than one third of the members of the American Management Association (AMA), the nation's largest management development and training organization, tape phone conversations, videotape employees, review voice mail, and check computer files and E-mail, a recent AMA report states. Scrutiny of job applicants has intensified, and this has fueled a boom in companies that do database searches of applicants' credit reports, driving and court records, and even workers' compensation claims. Personal behavior is no longer off limits: Some firms have adopted rules that limit co-workers' dating. Others ban off-the-clock smoking and drinking. Many companies regularly test for drugs.

Much of this occurs without the workers' knowledge. While companies say they collect information on their employees to comply with the law and protect their business interests, a recent survey of *Fortune* 500 companies showed that nearly half collect data on their workers without informing them. A majority said they share employee data with prospective creditors, landlords, and charities. "Many of these invasions—unthinkable a few years ago—have become institutionalized," says Craig Cornish, co-chair of the American Bar Association's workplace-privacy group.

Why has the worker's sphere of privacy shrunk? Employers say they feel intense pressure from lawsuits of every sort. The number of sex, race, disability, and age-discrimination suits brought by workers has more than doubled from over 10,700 in 1992 to 23,000 in 1996. Some of the cases focus not on management misbehavior but on what employees do to each other. For instance, workers assaulted by co-workers have sued their employers for negligent hiring. Morgan Stanley, a big Wall Street brokerage, was sued for $70 million by workers over racist jokes that appeared on the company's E-mail system. The plaintiffs claimed the jokes created a hostile work environment. The case was dismissed. "Nothing would please my clients more than to never again read an employee's E-mail," says Jay Waks, a New York attorney who represents corporate clients in employment litigation. Still, he says employers have no option but to protect themselves: "If they're going to be held liable, they'd better monitor."

Skyrocketing medical costs provide another incentive to probe, especially for companies that self-insure. Denise Nagel, the founder of the National Coalition for Patient Rights, says insurers have told her of employers that routinely check into workers' medical records. Nagel says they tell her "it happens all the time." Meanwhile, technology—particularly new software that can track and record everything workers do on their computers—is making it easier to peek over a worker's shoulder. And in most places, it's quite legal. State laws vary widely, and few federal statutes address workplace privacy. The FBI needs a court order to tap a phone line, for instance, whereas employers have far more freedom to listen in.

Employee Pitfalls

While the benefits for employers are undeniable, there are obvious hazards for workers. Careers may be damaged when investigators overreach, when mistakes are made, or when managers are too aggressive in enforcing company rules. To thwart sexual harassment charges, which have risen from 6,100 in 1990 to 15,300 in 1996, many companies have adopted "antifraternization," or antidating, policies. Wal-Mart is one such company. The giant retailer says the policies were also adopted to prevent favoritism, either real or perceived. Still, Joe and Tiffany Peters, of Dodge City, Kansas, got snagged in Wal-Mart's attempt to pre-empt problems. The newlyweds, both 24, met a year ago while working at the local Wal-Mart. Joe, who was then an assistant manager, says he followed company rules, told his boss he wanted to date Tiffany, and got an OK. Since Wal-Mart requires one member of a couple to quit or transfer, Tiffany gave two weeks' notice to leave.

But on Tiffany's last day at work, she says, two district managers grilled her for two hours. She was reduced to tears, she says. "They asked explicit questions. They wanted me to say we'd had sex," she claims. "The woman kept asking over and over: 'What kind of sex was it?' It was the worst thing that ever happened to me." Wal-Mart spokesperson Daphne Davis contests Tiffany's account. Davis would not say what questions the managers asked—only that they were necessary to determine whether Wal-Mart's policy was violated.

Nothing Personal

More companies are drug testing, too, hoping it will lead to improved worker performance and lower medical costs. In 1987, only one fifth of corporations drug tested workers, usually when drug use was suspected. But according to a 1996 survey by the American Management Association, more than 80 percent of corporations drug test workers, usually at random. Similarly, drug testing of job applicants is now commonplace. "If you had told people 20 years ago they'd have to drop their pants and pee in a jar to get a job, they'd have thought you were crazy," says Louis Maltby, director of the American Civil Liberties Union's (ACLU) workplace-rights office.

Dilbert © 1998 United Feature Syndicate, Inc. Reprinted with permission.

What worries some about drug tests is their reputed inaccuracy. John P. Morgan, professor of pharmacology at the City University of New York Medical School, says the false-positive rate on a typical unconfirmed drug test is 10 to 15 percent. Evelene Stein says hers is one such case. For years, Stein submitted to random urine tests given by the Crowne Plaza Nashville, where she was banquet captain. Then three years ago, the 53-year-old grandmother tested positive for drugs—the company would not tell her which drugs—and was fired. Stein believes the lab mishandled her urine sample. To try to prove her innocence, Stein says, she offered to pay for another test. The company refused, and many months passed before she found another job. "They ruined my reputation," says Stein. The Tennessee Supreme Court ruled the state's constitutional guarantee of privacy did not apply to private employers. Ralph Tipton, a spokesman for the hotel, declined to comment on Stein's case. He says the hotel policy "gives employees a safer, more productive place to work."

Some of the greatest strides in worker monitoring have come through technological advances. Nurses in over 200 hospitals now wear badges connected to infrared sensors to track their whereabouts. One badge manufacturer, Executone Information Systems, says they are designed to help route nurses to the patients who need them. The Tropicana Casino in Atlantic City is testing an infrared detection system that alerts the boss when workers leave the restroom without washing their hands.

Electronic-mail monitoring is even easier. In 1993, a *Macworld* study showed that 9 percent of companies searched employee E-mail. In 1996, a survey by the Society for Human Resource Management found that 36.4 percent of respondents search employee E-mail for business necessity or security. More than 70 percent said an employer should reserve the right to read anything in the company's electronic-communications system—while just one third have an E-mail policy. "The number of corporations who monitor E-mail but don't tell employees is appalling," says Beth Givens, director of the Privacy Rights Clearinghouse in San Diego and author of *The Privacy Rights Handbook*.

Flagging Phrases

Such practices have fueled the growth in the worker-monitoring business. One of the latest products is called Assentor. The software, created by SRA International, uses language patterns to allow corporations to search workers' E-mail for more than just simple key words. Financial-securities firms are testing the software to finger brokers who trade insider information, by flagging phrases like "hot little tech stock." The software—geared to comply with a Securities and Exchange Commission proposal that would require the monitoring of E-mail between stockbrokers and their customers—was made available in 1997. SRA plans to develop similar programs for the banking, law, and health care industries.

At other places, Big Brother is literally watching. Salem State College in Massachusetts had installed a video camera in an office, it says, for security reasons. But the college didn't tell its workers. Gail Nelson, a secretary, says she was particularly embarrassed by the videotaping because she often changed out of her work clothes after she closed the office for the day. Nelson complained and has hired an attorney, even though her state has no statute specifically prohibiting such videotaping. "Years ago, the law didn't recognize a wife could be raped by her husband," says Nelson. "But that didn't mean wives weren't raped by their husbands."

Employees may think their privacy is protected by the Fourth Amendment, which protects against unreasonable searches and seizures. But courts have ruled that the Constitution, which offers some protection to government workers, doesn't apply to employees of private firms. And while some states have passed privacy legislation, the protections are scattershot. North Carolina, for example, prohibits the tapping of telephone lines, yet allows employers to test for HIV in their annual physicals. Vermont prohibits HIV testing as a condition of employment but has no law against testing for genetic diseases. Few states have legislation to protect workers from being secretly videotaped or from employers' reading their E-mail.

Even so, workers have gained some new protections. Beginning in October 1997, the Fair Credit Reporting Act required employers to get written consent from job applicants before requesting their credit history. The United States Court of Appeals for the 10th Circuit in Denver upheld a lower court's decision that said employers cannot, under normal circumstances, force workers to disclose the prescription drugs they take. Finally, then President Bill Clinton said in 1997 he intended to offer legislation that would forbid disclosure of genetic information to employers and insurers. Several similar bills [have been considered by] Congress.

While workplace-privacy experts support the ban on genetic-information disclosure, they fear that if such legislation passes, politicians will then consider the entire issue resolved. And the far more common, yet less dramatic, encroachments on worker privacy will continue.

"In order to balance the employer's interests with those of the work force, the employer should offer a means by which the employee can control the monitoring to create personal boundaries."

Workplace Monitoring Can Be Ethical

Laura Pincus Hartman

Whether employees have a right to privacy has become a controversial question with the recently widespread use of personal computers and e-mail in the workplace. While some employers maintain that productivity is heightened by the monitoring of employees, others contend that video cameras, Internet tracking, and e-mail reading violate employees' rights to privacy. In the following viewpoint, Laura Pincus Hartman claims that employers may monitor employees without violating employees' privacy by creating a policy and ensuring that all employees are informed. Hartman is the director of the Institute for Business and Professional Ethics, whose mission is the teaching and training of ethical behavior to professionals and the public.

As you read, consider the following questions:
1. According to Hartman, why may an employer be justified in monitoring an employee's e-mail?
2. As cited by the author, what evidence did Bruno Frey discover about employee monitoring?
3. How is unreasonable intrusion into privacy defined by law, as cited by the author?

Reprinted, with permission, from "The Rights and Wrongs of Workplace Snooping," by Laura Pincus Hartman, unknown date, taken from the Institute for Business and Professional Ethics website at www.depaul.edu/ethics/monitor.html.

Employers have a number of reasons to monitor email—more than two million, if you ask Chevron Corporation. Recently, Chevron was required to pay four plaintiffs a total of $2.2 million where plaintiffs' attorneys found email evidence of sexual harassment. The attorneys had found the "smoking gun" when they located, on Chevron's email server, an email message that was sent to a number of people within the firm containing a list of jokes about "why beer is better than women." Had Chevron been monitoring its employees email, it may have seen the problem coming.

On the other hand, firms have been legally chastised for monitoring email, as well. Recently, the Boston Sheraton settled a lawsuit brought by its employees for more than $200,000. The employees claimed an invasion of privacy where the hotel had secretly monitored the hotel's employee locker room.

Most employers believe that they know what their employees are doing when they are in their offices. Working! They hope. . . . Does one have a right to know what they are doing? Considering recent increases in employee salaries and corporate liability for employee actions, one might expect you to be interested, at least. On the other hand, given the high expectations about employee loyalty and commitment in terms of hours spent at their workplaces, one might expect that an employee might conduct a bit of personal business during the course of the work day.

The balance to this issue is challenging, especially in light of recent advances in technology that allow for intrusions into an employee's personal life in ways never before imagined. While the ease of access has been impacted by technological advance, an employer's interest in personal information is nothing new. Almost a century ago, Henry Ford used to discover the morals and hygiene of his assembly line workers by sending investigators into the community. Other managers would have access to their employees' personal information simply because everyone lived in the company town.

Is There Any Workplace Privacy Anymore?

Today, invasions of privacy in the workplace occur far more frequently than one might expect. In fact, while a 1993 study indicated that 30% of 1,000 firms surveyed had searched their employees' computer files, electronic mail, and voicemail, subjecting more than 20 million employees to computer monitoring alone, a more recent study evidences the explosion of growth in this particular area. A survey conducted and released in May 1997 by the American Management Association revealed that 63% of mid-sized to large firms conducts some form of electronic surveillance. . . .

A supervisor may have completely justifiable reasons for considering this type of monitoring or for evaluating an employee based on these criteria. Consider an employer's interest in employee personal information or emails in order to ensure compliance with discrimination laws, to administer workplace benefits, or to appropriately place workers in positions. Recall Chevron's experience with the gender-based email jokes. Had Chevron been in the practice of monitoring email messages, and had the employees known of this practice, their resulting liability may have been far less likely. Instead, employees would have been forewarned that only business-related messages should be sent using the firm's internal mail system, thus saving Chevron from defending the actions of its employees.

In addition, the more complicated a task, the more necessary effective workplace supervision becomes. Instead of editing a document by using interoffice mail to transmit it from the author to her or his supervisor, the supervisor may choose to save time by simply reading it on the author's computer. On the other hand, workers feel a lack of respect from their employers looking over their shoulder at every turn, which may in turn affect productivity or the culture of the workplace.

No Legal Protection

While some workers believe that they are safe from such intrusions or express no concern about the sharing of their personal information, reports of intrusion horror stories abound. James Russell Wiggins' employer conducted a background check and fired him because the report showed a prior conviction for cocaine possession. Despite Wiggins' protests that the information was patently false, his company refused to rehire him. Later it was discovered that his identity had been confused with that of James Ray Wiggins, and a lengthy lawsuit ensued.

Indeed, according to a congressional report, half of all credit reports and background checks contain mistakes. The American worker is becoming more aware of the possibility for intrusions or violation, as well. A survey conducted by Louis Harris & Associates and Dr. Alan Westin showed that 89% of the American public is concerned about threats to their personal privacy, with 55.5% saying that they are "very concerned."

While no related case has yet reached the Supreme Court, these actions have received lower court attention. As early as 1990, Epson America survived a lawsuit filed by a terminated employee who had complained about their practice of reading all employee email. The court found in favor of Epson because the employees were notified that their email might be monitored. However, relying on court precedent for protection is a double-edged sword. An employee-plaintiff in one federal action won a case against his employer where the employer had monitored the worker's telephone for a period of 24 hours in order to determine whether the worker was planning a robbery. The court held that the company had gone too far and had insufficient evidence to support its claims. In another action, Northern Telecom settled a claim brought by employees who were allegedly secretly monitored over a 13-year period. In this case, Northern Telecom agreed to pay $50,000 to individual plaintiffs and $125,000 for attorneys' fees.

What Is Technologically Possible?

Technology may affect workplace privacy through a variety of mechanisms. Undisclosed or disclosed monitoring of employees has reached new proportions with the ease and availability of clandestine monitors. Counter Spy Shop has retail outlets in several large cities and specializes in high technology gear for monitoring purposes. The firm, which does business on the internet as well as through a traditional sales force, sells devices that allow firms to conduct covert audio and video surveillance, in addition to items that can encrypt or scramble any of the firm's transmissions (to ward off potential corporate saboteurs). According to Spy Shop's sales manager Tom Felice, "the more discreet they are, the more popular." One-third of the *Fortune* 500 has been a client of the Spy Shop. The Spy Shop sells not only traditional monitoring equipment, but also such scientifically questionable devices as one that tells the caller whether the individual on the other end of the line is telling the truth. The device uses voice stress analysis of voice tremors in order to determine whether the individual is lying and costs about $5,000.

Creating an Employee Policy

Because companies often rely on network administrators and other information technology personnel to retrieve or intercept employee e-mail, it is critical that information technology managers understand the scope of their company's right to monitor employee's e-mail and corresponding privacy rights of employees. This task is complicated by the limited and ambiguous nature of current laws concerning the privacy of electronic communications. Few cases interpreting those laws exist, and the decisions in those cases are inconsistent, giving little guidance to companies. But a review of the current statutes and cases in this area clearly indicates that, at a minimum, an employer's best defense against employee privacy claims is a comprehensive policy governing the monitoring of electronic communications and computer files.

H. Thomas Davis, *Network Magazine*, February 1, 2001.

Other, far less traditional methods of monitoring now exist as a result of technological advances. One product, called the Truth Phone, promises to analyze voices during telephone calls in order to detect possible deception. New Jersey based Net/Tech offers a product called Hygiene Guard that tracks whether a company's employees are using soap dispensers and washing their hands after they use the restroom. If they fail to do so, the device may beep periodically or flash to alert supervisors. "They're starting with these little badges. The next thing, they're using video cameras. Some people feel violated. It's an insult," claims a union employee. The local union in one case is concerned that the purpose of the badges is not to ensure clean hands, but to protect against workers lingering too long or who make too many trips to the restrooms.

Because many employees use personal computers on the job, and because those computers are often linked either to the Internet or, at least, to an internal network, monitoring employees has become simpler. But the reasons employers have for monitoring have become more complex. For example, assume that ABC Corporation employees repeatedly access specific locations of competitors' web sites looking for competitive information. By tracking these hits, ABC's competitors might learn which of their technology interests ABC's employees, and this may give those competitors insight into the direction of ABC's research and development. Further, firms have reason for concern if employees download program files without compensating the creator or use copywritten information from the web without giving credit to the original author. These actions can expose the firm to potentially significant copyright infringement liability. Finally, Internet access makes a company vulnerable not only to unauthorized access by hackers but also to numerous viruses, which employees can inadvertently introduce to company systems by downloading software programs from the web or even simply exchanging email with others outside the company.

In fact, email raises a host of additional questions. Monitoring these transmissions is becoming more common in corporate America and elementary to even the most basic technician. Employers may monitor email transmission to make sure their trade secrets remain secret or to ensure that email is used only for business purposes. They also may want to maintain consistent quality of everything that goes out under the firm's "letterhead." In other words, when an employee sends an email using a company-provided email server, usually the firm's name is identifiable through the individual's email address. A firm should be just as concerned about what goes out above that email signature as it is about what goes out on its letterhead in order to ensure that inappropriate communications are not considered "employer authorized" in a legal context.

The Criticisms

While no one would express a privacy concern about a manager's choice to read over a business letter to a client before an employee mails the letter, drawing the line between that which is personal and that which is public is difficult.

Employers argue that monitoring is an effective means to ensure a safe and secure working environment and to protect individuals and the firm's assets or resources. In addition, some contend that monitoring may boost efficiency, productivity, and customer service, and allows them to more accurately evaluate performance. For example, monitoring is considered to be a "real time" aid to performance appraisals. Supervisors have easy, immediate access to information that will help them in their nurturing and evaluation of their employees. In addition, a manager is now able to see or read exactly what an employee does during the day, rather than relying on second-hand reports. The manager is thereby more able to review the employee's performance; consider the impact of telephone monitoring on the evaluation of a customer service resspresentative. The supervisor who monitors telephone calls to the service desk can now truly review her or his employees at work.

On the other hand, critics of monitoring point to research evidencing a link between monitoring and psychological and physical health problems, increased boredom, high tension, extreme anxiety, depression, anger, severe fatigue, and musculoskeletal problems. In his 1992 research, Swiss economist Bruno Frey found evidence that monitoring worsened employee morale and thereby negatively affected their performance. This was primarily the result of the employees feeling like the employer had low expectations of them (because the firm felt the need to monitor), so the employee, in essence, lived down to those expectations! In addition, critics of monitoring are concerned with the employee's legitimate expectation of privacy in certain areas of the working environment, and whether the employees are notified of the presence of monitoring.

Under common law, *unreasonable* intrusions into employees' private affairs are prohibited. An invasion of privacy, called "intrusion into seclusion," is defined as "intentional intrusion upon the solitude or seclusion of another that is highly offensive to a reasonable person." Private sector employees, therefore, have privacy in those areas in which they have a legitimate expectation of privacy—a restroom or changing room, for instance. In *K-Mart v. Trotti*, the court held that the search of an employee's company-owned locker was an unlawful invasion of privacy since the employees used their own locks and therefore had a reasonable expectation of privacy. On the other hand, the law is murky on this issue; for example, an employer's search of employee lunch buckets was held reasonable by another court.

Employees obviously would have no reasonable expectation of privacy where an employer *notifies* them that they will be monitored in specific situations during specific times. But notification does not provide complete immunity from charges of invasion of privacy. A Kansas District Court remarked that "a reasonable person could find it highly offensive that an employer records an employee's personal phone calls in the circumstances where the employer did not discourage employees from making personal calls at their desks and did not inform the plaintiff employees that their personal calls would be recorded." When applied to email, this finding suggests that an employee has a reasonable expectation of privacy if an employer issues an employee a password or suggests that email is confidential.

Ethical Concerns

What is the *ethical* answer? In the case of employee privacy relating to monitoring, the employer must make a decision about how to handle her or his need to supervise and to evaluate workers. The first step of the process is to determine the values of the firm. This step may already have been taken if the firm has developed a mission or statement of principles. Next, the employer must consider whether monitoring satisfies the goals or mission of the firm. Assuming it does (since a negative relationship here would end the discussion and resolve the dilemma), the employer must be accountable to those affected by the decision to monitor by considering their personal interests.

In order to respect the privacy rights of the employees and

their right to make informed decisions about their actions, the employer should give adequate notice of the intent to monitor, including the form of monitoring, its frequency, and the purpose of the monitoring. In addition, in order to balance the employer's interests with those of the work force, the employer should offer a means by which the employee can control the monitoring to create personal boundaries. In other words, if the employer is randomly monitoring telephone calls, there should be a notification device such as a beep whenever monitoring is taking place *or* the employee should have the ability to block any monitoring during personal calls.

A monitoring program that is developed according to these strictures not only respects the personal autonomy of the individual worker, but also allows the employer to supervise effectively the work done, to protect against misuse of resources, and provides an appropriate mechanism by which to evaluate each worker's performance, respecting the legitimate business interest of the employer.

Periodical Bibliography

The following articles have been selected to supplement the diverse views presented in this chapter. Addresses are provided for periodicals not indexed in the *Readers' Guide to Periodical Literature*, the *Alternative Press Index*, the *Social Sciences Index*, or the *Index to Legal Periodicals and Books*.

Arthur Allen	"Exposed," *Washington Post*, February 8, 1998.
Michael J. Blotzer	"Privacy in the Digital Age," *Occupational Hazards*, July 2000.
Christian Science Monitor	"Will Privacy Rights Pass the Smell Test?" *Christian Science Monitor*, August 23, 2000.
Richard Dalton	"The All-Seeing, All-Hearing Monster," *Byte.com*, April 28, 2000.
Bob Evans	"Online Privacy: Protect It or Lose It," *InformationWeek*, February 28, 2000.
Simson Garfinkel	"Privacy and the New Technology: What They Do Know Can Hurt You," *Nation*, February 28, 2000.
Aimee Howd	"Medical Records Are Up for Grabs," *Insight On the News*, March 15, 1999.
Robert Kuttner	"The Age of Trespass," *American Prospect*, January 1999.
Toby Lester	"The Reinvention of Privacy," *Atlantic Monthly*, March 2001.
Patrick McCormick	"Keep Your Eyes to Yourself," *U.S. Catholic*, March 2000.
Monica Rogers	"Does Snooping on Surfers Invade Rights to Privacy?" *Crain's Chicago Business*, May 15, 2000.
William Safire	"Living in the Age of Surveillance," *New York Times*, March 13, 2001.
San Francisco Chronicle	"The Eyes of Technology," *San Francisco Chronicle*, March 14, 1999.
Charlie Schmidt	"Beyond the Barcode," *Technology Review*, March 2001.
Michelle Singletary	"Whose Information Is It Anyway?" *Washington Post*, January 31, 1999.
David Wessel	"Privacy vs. Productivity: A Tough Choice," *Wall Street Journal*, March 8, 2001.

CHAPTER 4

How Will Technology Affect Society in the Future?

Chapter Preface

The advent of the twenty-first century has given rise to various forecasts of technological evolution. Innovations in genetic engineering, robotics, and nanotechnology—a field of science that strives to control individual atoms and molecules to create computer chips and other extremely small devices—have spawned questions surrounding the ethics and potential consequences of society's increasing dependence on technology.

Many argue that technological advances in robotics will threaten the future of humans by creating a superior form of life that is able to self-replicate. Hans Moravec, a leading robotics researcher, argues that robots created with the ability to reproduce will also be endowed with survival instincts that may jeopardize the human race. In his book *Robot: Mere Machine to Transcendent Mind*, he writes, "Robotic industries would compete vigorously among themselves for matter, energy, and space, incidentally driving their price beyond human reach. Unable to afford the necessities of life, biological humans would be squeezed out of existence." Moravec and others contend that by creating a form of life more intelligent and physically superior to humans, people risk the endangerment and extinction they have inflicted on countless species.

Others maintain that because robots are not alive, do not have biological needs and desires, and have no need to compete, their existence will not threaten humans. Chris Malcolm, a lecturer on artificial intelligence, argues that unless humans program competitive drives in robots, they pose no threat to people: "They will *not* be animals, they will *not* have a competitive instinct, they will *not* have an instinct to reproduce themselves in competition with us, they will *not* compete with us for dinner, oil, or electricity. To be very simple about it, they will not object to being switched off." Supporters of Malcolm's argument claim that while intelligence can be programmed, biological instinct cannot.

The consequences of artificial intelligence is one of the issues debated in the following chapter on the future of technology.

"Considering the interactions between technology and society, we can . . . foresee many potential consequences on business and social life."

The Future of Technology Can Be Predicted

Ian D. Pearson

While many argue that present conditions are not necessarily indicative of the future, Ian D. Pearson maintains in the following viewpoint that predictions about the future can be accurately made. By extrapolating the progress rates of technologies currently being researched and developed, Pearson predicts that by 2020, humans will live longer and be more healthy, alternatives to fossil fuels will provide cleaner energy, and incredibly fast computers and robots will be commonplace. Pearson is a telecommunications analyst and futurologist in Great Britain.

As you read, consider the following questions:

1. According to the author, what possibility does a complete map of the human genome provide?
2. How will telecommunications benefit the elderly, according to the author?
3. How will the computers of the future differ from those of the present, as predicted by Pearson?

Reprinted, with permission, from "The Next Twenty Years in Technology: Timeline and Commentary," by Ian D. Pearson, *The Futurist*, January 2000. Copyright © 2000 by the World Future Society.

No one knows precisely what life will be like in 2020, but if we know the development rates of different technologies, we can anticipate many of the things that will be possible and when they are likely to happen. Considering the interactions between technology and society, we can also foresee many potential consequences on business and social life. Such scenarios allow us to plan with a much better idea of how life might look, realizing that many things will still turn out differently in spite of our best efforts.

Computer-based technology will change the most. By 2020, we will share our planet with synthetic intelligent life-forms; they may even have legal rights. Overall, they will catch up with human intelligence before 2020, though there will still be a few things that only people can do. Most new knowledge will be developed by synthetic intelligence; we won't understand some of it, even as we accept the benefits. As machines gradually take over both mental and physical work, we will shift to a "care economy," where people gradually concentrate more on the human side of activity. Partnership between man and machine will make our work more productive and our play more enjoyable. Even entertainment will be within the machine domain, with today's crude computer game heroes and heroines evolving into a whole range of entertainers, even chat-show hosts. It is even possible that some of our friends may be synthetic, and, since many of our relationships will be Internet-based, we won't necessarily know which ones.

Longer Lives and Digital Immortality

Medical technology is improving rapidly, too, and by 2020 new babies can expect to live well over 100 years and perhaps to 130. In fact, for a while, life expectancy will increase faster than people get older. By the time today's babies reach the limits of their life expectancy we will probably be able to download their minds into electronic storage. They will be able to carry on digitally even when their bodies are dead and buried. Imagine making a speech at your own funeral!

With a complete map of the human genome, we will be able to customize our children. This technology is quite disturbing to many people, and such developments are likely to be resisted in many countries. However, it is also likely that someone somewhere will offer such a service—at a price.

Our understanding of the body will be much greater, and we will be able to treat many conditions much more successfully than today. More than 95% of body weight is made up of parts that could be replaced by synthetic alternatives by 2020. Many organs, and perhaps even limbs, could be replaced by fully organic replacements grown in the lab.

Information technology will help enormously. Apart from having a full multimedia medical record, including scans and videos of past operations, we will perhaps routinely wear health monitors. Our electronic environment will react to our emotional and physical state automatically, reducing stress. By linking directly to our nervous system, computers could pick up what we feel, and perhaps even stimulate feeling, too, so that we could develop full sensory virtual environments. This will be the beginning of the long process of man-machine convergence that will culminate in a fully electronic human well before the end of the next century.

Teleworking and Teleteaching

Improving technology will revolutionize the way we do business and earn our living. Only a few people will be needed to staff our agricultural and manufacturing industries, and most of today's service industries will be automated to a large degree. People will focus much more on interpersonal roles, the human side of work. Most companies will bring in staff on a project-by-project basis. This staff will use advanced communications to work together from anywhere as if they were in the same office. Workers will change jobs frequently, but won't want to move each time, so they will most likely make use of abundant telework centers, full of infotech-equipped hot desks. Local communities will benefit from reduced commuting, lower stress, and greater cohesiveness.

On the government side of business, we will see a push for global taxation so that information companies can't avoid tax just by relocating their software around the Net every few seconds. Business transactions will all be automatically monitored, and taxation will be fully integrated into funds transfers.

Education will not be confined to a single school or university. Students will attend key lectures via the Net, or learn by experience, interacting with simulations in advanced computer systems. Some teachers have that special gift, and we can expect many superstar teachers with huge followings around the world.

Education often overlaps with leisure and entertainment, so the use of virtual environments will be widespread by 2020. Even today, a giant "hamster-ball" mounted on an air bearing allows a person to wander around a computer-generated space projected onto the outside of the ball. With "active" contact lenses that can input images directly into the eye, which we expect around 2010, virtual reality could be both high resolution and fully three-dimensional. With a direct link into the nervous system to create synthetic sensation, we will have all we need to produce a "holodeck" like the ones in Star Trek.

Exploring Virtual Environments

Virtual environments may be created for entertainment, sports, education, shopping, and even business meetings. We are only limited by human imagination, and even that won't be a limit by 2020: Computers will design new places for us to explore. Virtual environments will greatly change the ways we do things. We should hope, however, that virtual escapism doesn't become a social problem.

Demographically, our society will be much older, with many more retired people. Younger people will be highly taxed to pay for older people's pensions. This might cause conflict between the generations. Younger people might make use of teleworking technology to reduce their local tax bills through electronic emigration; they could also avoid taxes by taking remuneration in information products or entertainment instead of money. They might vote with their feet and actually emigrate to countries where the population is younger.

Loneliness will be much less of a problem for old people as they can keep in touch with friends and relatives via large screens with life-size images. They can also make new friends with similar interests easily on the Net. Furthermore, ongoing automation of most work is likely to cause people to redefine their identity in terms of social involvement rather than through their job; this trend will benefit the whole community. Not all communities will be geographically based: People may belong to large Internet communities, too. These virtual communities will have significant political power and the ability to wield it on very short notice.

Cleaner Energy and Environment

All these new technologies will need power, and developments in solar power will likely provide it. Transportation might rely on catalytic solar breakdown of water to provide hydrogen for fuel cells. Other forms of renewable energy will be much further developed, too, but we still won't have nuclear fusion by 2020. The environment will start to improve as we reduce fossil fuel use. I expect that we will see a growth of scientific environmentalism: People will expect more professional studies of problems rather than the emotional reactions that have sometimes proved counterproductive.

Predicting the Future

On a flight from Tokyo recently, my gaze wandered from the computer sitting in my lap to the Pacific 20,000 feet below. As we descended toward Los Angeles, I watched the blue ripples relentlessly marching toward a distant shore. Suddenly I realized it would be possible to calculate from the waves' direction and velocity not only where they would end up but when they would get there. In a sense, I could predict their future. Turning back to my computer, I marveled at how quickly it had become an essential appendage, and it occurred to me that the waves below were not all that different from the waves of technological change that sweep over us. Like waves at sea, waves of technology are always out there somewhere, relentlessly heading for an impact. And they can be detected while still far from shore. We even ought to be able to see them coming when they're 20 years away.

Eric Haseltine, *Discover*, October 2000.

Many people will greatly mistrust megacorporations and choose to produce their own food. Farming cooperatives will outsource people's vegetable plots, producing food according to whatever the individual consumer demands.

Around 2020, sensors may be deployed in the countryside to monitor all manner of things, from climate to insect population. If insect pollination has suffered from greater pesticide use or increased use of genetically modified crops, then we may see widespread use of robotic insects to do this job.

At home, gadget lovers will have digital bathroom mirrors, wristwatch cameras, virtual fish tanks, and electronic paintings set against electronic wallpaper that adjusts to the mood of the inhabitants and reduces background noise. And that is just within the next decade. Soon after that, intelligent gadgets that anticipate what you want—and often get it wrong—could bring about cases of "kitchen rage" and a new breed of technician: the robotic psychiatrist. Robotic psychiatrists would be humans who diagnose and treat malfunctioning robots or computer programs. Most gadgets will be able to talk to their psychiatrist, and you, through interactive voice response.

Screens might replace windows in less-attractive neighborhoods, giving occupants the impression that they lived somewhere much nicer, and of course the location could be different each day. By 2020, we may see common use of ultra-high technology to mask the apparent impact of technology, digitally enhancing our environment and reducing technological intrusion, The most traditional pub might be one that uses digital windows so that you can watch the horses and carts outside and experience the noises and smells of times gone by.

Life with Machines

Full voice interaction with machines and other language applications such as translation have been promised for some time and always seem just on the horizon. Nevertheless, by 2020 we can expect to converse with a smart machine about anything, and it will usually have the intelligence to implement appropriate requests. Computers are likely to have faces and personalities. Voice synthesis should sound perfectly natural.

The computers of 2020 won't look like those today; most of the time we won't see them. They will be ubiquitous but invisible, hidden in infrastructure and in almost every device around us. They are likely to be at least a hundred thousand times faster than today's computers, maybe even a million times faster. Memory for computers will be in the million gigabyte range, more than the human brain. Storage will no longer be based on disks. Moving parts will disappear, replaced by holographic and a variety of other forms of storage. We will still need supercomputers, maybe to run a company's board of directors, and some of these may have a billion (1,000 million) processors: some digital, some analog, some neural, some quantum, and some using molecular computing techniques.

Some of these computers will be mobile—i.e., robots. They will have an array of sensors comparable or superior to anything in the natural world. In some countries, robot population will approach human population by 2020. We will use robots for all manner of jobs around the office and home as well as in factories and agriculture. Millions of robots will be small and insectlike. There may be a flock of them maintaining the lawn and keeping the garden tidy, picking up individual leaves and carrying them to the compost heap. Some robots will act as pets and toys; others will take part in sports.

We will not have androids like Star Trek's Data by 2020, but many robots will have some organic characteristics. Some might use muscles based on gels that resemble organic muscles. And imagine a robotic garden gnome that catches the occasional fish in the garden pond, albeit a robotic fish.

Cyberspace will increasingly affect every area of our lives, with 75% of the world's population using the Internet by 2020. We will take access to the Net for granted wherever we are. Cordless communication will be the norm long before then, but the infrastructure will still be fiber. However, capacity will be a problem even on optical fiber. Demand is increasing very fast today and is likely to accelerate in the future, so we will one day approach the maximum capacity of fiber—something we considered infinite not so long ago.

We will wear a multitude of devices, not just watches and phones. Jewelry that reacts to our emotional state, audio and video players, translators, and cameras built into our glasses are just a few examples. We might even have displays built into our clothes, but fashion is impossible to predict.

Antitechnology Backlash

With large, Net-based communities crystallizing around commonly shared values, we can expect trouble, even occasional cyberwarfare disturbing the Net. Security measures may be much more pervasive. In fact, with video cameras everywhere linked to automated recognition systems, and with all electronic transactions potentially monitored, people may be watched in everything they do. Such a high degree of electronic intrusion, coupled with many jobs being automated, could possibly lead to a backlash against all the new technology. Large numbers of people may form a parallel society using lower technology and aiming for a more traditional lifestyle.

Money will be mainly electronic, and there will be at least one global electronic currency that can be used on any Web site, or in any shop, anywhere in the developed world. As the Net gradually becomes the standard platform for most commercial activity, this global currency will soon become the currency of choice, causing the dollar, euro, and yen to evaporate into oblivion. We could of course have electronic dollars and euros and yen alongside a global electronic currency, but why bother? The equivalent today would be being paid in store vouchers—people much prefer currency that they can use anywhere, and on the Net we will want to shop globally most of the time. In lieu of electronic cash, barter systems may arise in some Net communities and local exchange trading systems.

In a world of strong electronic signatures, encryption, and integrated systems, there may no longer be any need to put money in a bank, and many people may choose to keep control of it themselves. Banks will be forced to add many new services or go out of business.

As for human transport, we will need a fully integrated road traffic information and management system to cope with increased automobile traffic. Such systems can double the capacity of a road while reducing the stress of travel. Long before we leave, a computer might negotiate the route to guarantee we arrive at our destination on time, and it will automatically reroute us on the way if there's a problem. All we will have to do is sit back and watch the scenery as the car automatically takes us where we want to go. Many of our vehicles will be powered by nonpolluting hydrogen fuel cells. However, with world travelers—including over 5 billion air travelers—racking up 50 trillion kilometers in 2020, some popular destinations may react to overuse by severely limiting the number of physical visitors, forcing the remainder to travel virtually on the Net.

The future looks different in many ways, but some things won't change. We will still have the same fundamental needs and desires as the cavemen: food and shelter, love, status, and self-fulfillment. And we will still squabble sometimes. These human attributes are written in our DNA, and while we might begin to tinker with that, some of our nature goes very deep indeed.

> "*Real world events can rarely be replicated to test predictions.*"

The Future of Technology Cannot Be Predicted

Roger A. Pielke Jr. and Dan Sarewitz

The dawn of the twenty-first century has pundits racing to predict changes in societal conditions and in people's lives. Recent technological advances, some argue, promise further economic growth and herald the potential for robots and cloned humans. Others claim that variables such as natural phenomena and human nature make accurate predictions of the future impossible. In the following viewpoint, Roger A. Pielke Jr. and Dan Sarewitz contend that predictions about the future must be regarded with skepticism, as most scientific experiments cannot be replicated in reality. Pielke is a scientist at the Environment and Social Impacts Group at the National Center for Atmospheric Research in Boulder, Colorado, and Sarewitz works as a researcher at the Center for Policy Outcomes at Columbia University and Georgia Tech.

As you read, consider the following questions:
1. How is prediction central to the scientific method, according to Pielke Jr. and Sarewitz?
2. According to the authors, why are data no substitute for experience?
3. Why can predictions about the future be more dangerous than ignorance, according to the authors?

Excerpted from "Prediction: Science, Decision Making, and the Future of Nature," by Roger A. Pielke Jr. and Dan Sarewitz, a 2000 Web article published at www.islandpress.org/ecocompass/predictions.html. Reprinted by permission of Alexander Hoyt Associates as agents for Island Press.

Y2K came and went, with various predicted cataclysms nowhere to be found. Did clever human intervention—at a cost of more than $500 billion—change the predicted future by fixing the glitch ahead of time? Or were the forecasts of disaster simply overstated?

Modern society attempts to predict all manner of future events, from the course of hurricanes to oil prices to asteroid impacts—in order better to prepare for them. Advances in science and technology—especially the ability of computers to rapidly process massive amounts of data using sophisticated mathematical models—underlie these efforts to foretell the future. But, in a complex world, accurate predictions are often surprisingly hard to come by. Just prior to Y2K, for example, western Europe was ravaged by a storm that took weather forecasters by surprise, and caught tens of millions of people unprepared.

Prediction is central to the scientific method. Many scientific hypotheses are in fact predictions that can be tested under controlled experimental conditions. For example, a biologist might hypothesize that a certain gene controls appetite in rats. The biologist can test this idea by removing or altering this gene in laboratory rats, and seeing if the eating behavior changes. Once the prediction is confirmed, it can be applied to real-world problems, perhaps, for example, gene therapy for appetite suppression in humans.

But predictions about the real world are different from those that apply to experimental conditions. Real world events can rarely be replicated to test predictions. For example, we cannot replay the millennial transition, only this time without the half-trillion dollar intervention, to see if the Y2K bug would really have caused the predicted disruptions. Moreover, the conditions under which real world events occur cannot be controlled as they can in the laboratory. Unforeseen complexities often insinuate themselves, even under conditions that seem highly predictable. Consider the recent loss of NASA's Mars lander. The laws of physics allowed aerospace engineers to predict precisely the path of the rocket through space. But who could have predicted that the Mars mission would be doomed by the failure of engineers to make a simple yet crucial conversion between English and metric measurements? The physics predicted an absolutely certain future; human behavior intervened to confound the prediction.

Faulty Predictions

Humans, of course, are not the only source of uncertainty in predictions. A European storm in December 2000 shows how capricious nature can be. Similarly, scientists have been trying for decades to predict earthquakes, with little or no success. In the late 1980s, some of the world's best earthquake scientists predicted, with ninety-five percent certainty, that a major earthquake would strike a segment of the San Andreas Fault south of San Francisco within five years. The earthquake didn't occur. Is this a reflection of the five percent chance that there would be no earthquake? Or does it mean that the hypotheses on which the prediction was based were flawed? Recent research suggests the latter interpretation— the ninety-five percent certainty level was a vast overstatement of the actual probability. Which raises a key question: how can we assess the validity of predictions?

Can We Guess the Future?

Our knowledge of the future is always open to error and change, no matter how good the evidence and arguments for predictions seem to be. All predictions are a form of more or less informed guessing. Our best predicitons are projections into the future from what we know about the past and the present, and since the future does lie before us—it is not yet here—we cannot know that the future will be like the past.

Lloyd Eby, *World & I*, January 1, 2000.

Experience is the best guide. Every day, we read the weather forecast, and see the probabilities assigned to predictions of future events—eighty percent chance of rain, etc. The National Weather Service issues about 10 million weather forecasts each year, and this gives them lots of information that they can use to see how well they are doing. Of equal importance, it gives the public lots of opportunities to evaluate the predictions as well, and develop expectations that can guide actions. No one expects weather predictions to be perfect, and one's experience with past predictions allows one to make decisions in response to new predictions. For example, during the muggy summer months in Washington, DC, weather forecasts routinely show a chance of thunderstorms in the afternoon, but experience quickly teaches that this is not sufficient reason to cancel a picnic.

Data are no substitute for experience. It might seem logical that the more data and the more sophisticated mathematics that go into a prediction, the more accurate the prediction will be, but this is simply not the case. If a crucial piece of information is missing—say, that aerospace engineers will fail to make a conversion from metric to English measurements—then all the data in the world can't make up for it. But it is often impossible to know what the most important data are, or if they have been taken into account in making the prediction. Economic forecasts, for example, are supported by voluminous data processed through some of the world's fastest computers, but experience shows that they are often little better than random guesses. . . .

Assessing Strengths and Limitations

Understanding the strengths and limits of predictions is important because the actions we take today are based on how we view the future. If, for example, we are certain that things will turn out a particular way, then we may be justified in pouring our resources into activities based on this certainty. The Y2K bug is a good example of this type of prediction. But if, on the other hand, uncertainty is high—or if it cannot be evaluated—then we might be better off hedging our bets and preparing for a wide variety of possible outcomes. For example, predictions of global warming have focused international environmental efforts on reducing greenhouse gas emissions, even though the impacts of future warming remain uncertain, and even though society's growing vulnerability to climate is caused by many factors, not just greenhouse warming. At the same time, such problems as aging of the global population, migration of people to urban areas, spread of infectious diseases, and destruction of coastal ecosystems, offer interconnected challenges to human welfare that are arguably more certain than, and at least as serious as, global warming—yet they remain relatively neglected.

Predictions are often rooted in complex scientific theories and generated by advanced technological tools, so those who need to respond to predictions may find themselves at the mercy of experts. But the lay public can get a sense of the trustworthiness and value of predictions by asking some simple questions. Are the predictions testable in the laboratory? Are the predicted phenomena governed by well-known scientific laws, or do they encompass complex natural or societal systems? Has the predictive method been tested numerous times against reality, so that its reliability can be evaluated on the basis of direct experience?

Predictions of the future can be more dangerous than ignorance if they induce us to behave in ways that reduce our resilience in the face of inevitable uncertainties and surprises. Advances in science and technology mean that increasingly sophisticated predictions covering a widening variety of natural and cultural phenomena are sure to be available to society in the future. For society to benefit from these advances, it must always question predictions.

| "Few people have seriously considered being annihilated by a robot race."

Technology May Threaten Society

Will Knight

Recent improvements in computer technology have freed people from many tedious tasks, but in doing so have given much power and control to computers. Many claim that the wonders of technology may ultimately equip computers with too much intelligence and independence, which could give them a dangerous degree of control over humans. In the following viewpoint, Will Knight makes this argument and contends that if computers are created with too much intelligence and autonomy, they may attempt to eradicate the human race. Knight is a contributing editor at ZDNet, a website that provides assistance, information, and news about science and technology.

As you read, consider the following questions:

1. As quoted by the author, what is Isaac Asimov's First Law in regard to robotics?
2. What is Moore's Law, according to Knight?
3. What are the three stages people go through in examining future technology according to Ray Kurzweil, as quoted by the author?

Excerpted from "Intelligent Machines Threaten Humankind," by Will Knight, ZDNet, January 23, 2001, published in the News section of www.zdnet.co.uk. Reprinted with permission.

S cience fiction has portrayed machines capable of thinking and acting for themselves with a mixture of anticipa-
tion and dread, but what was once the realm of fiction has
now become the subject of serious debate for researchers
and writers.

Stanley Kubrick's groundbreaking science fiction film
2001: A Space Odyssey shows HAL, the computer aboard a
mission to Jupiter, deciding (itself) to do away with its human
copilots. Sci-fi blockbusters such as *The Terminator* and *The
Matrix* have continued the catastrophic theme portraying the
dawn of artificial intelligence as a disaster for humankind.

Science fiction writer Isaac Asimov anticipated a potential menace. He speculated that humans would have to
give intelligent machines fundamental rules in order to
protect themselves.

- A robot may not injure a human being or, through inaction, allow a human being to come to harm.
- A robot must obey orders given it by human beings except where such orders would conflict with the First Law.
- A robot must protect its own existence as long as such
  protection does not conflict with the First or Second Law.

Later Asimov added a further rule to combat a more sinister prospect: "A robot may not injure humanity, or,
through inaction, allow humanity to come to harm."

From Science Fiction to Reality

Will machines ever develop intelligence on a level that could
challenge humans? While this remains a contentious question, one thing is certain: computing power is set to increase
dramatically in coming decades. Moore's Law, which states
that processing power will double every 18 months, is set to
continue for at least the next ten years, and quantum computers, though poorly understood at present, promise to add
new tools to artificial intelligence that may bypass some of
the restrictions in conventional computing.

What was once the realm of science fiction has mutated
into serious debate. While the focus is currently on cloning
and genetic engineering, few people have seriously considered being annihilated by a robot race. . . .

Steve Grand, artificial intelligence researcher and author

of *Creation: Life and How to Make It*, says it would be impossible for humans to be totally sure that autonomous, intelligent machines would not threaten humans. Perhaps more worryingly, he claims it would be futile to try to build Asimov's laws into a robot.

Artificial intelligence researchers have long since abandoned hope of applying simplistic laws to protect humans from robots. Grand says that for real intelligence to develop, machines must have a degree of independence and be able to weigh up contradictions for themselves, breaking one rule to preserve another, which would not fit with Asimov's laws. He believes that conventional evolutionary pressures would determine whether machines become a threat to humans. They will only become dangerous if they are competing for survival, in terms of resources for example, and can match humans' intellectual evolutionary prowess.

"Whether they are a threat rests on whether they are going to be smarter than us," he says. "The way I see it, we're just adding a couple more species."

Eliminate Humans?

In his book *The End of the World: The Science and Ethics of Human Extinction* John Leslie, professor of philosophy at Guelph University in Canada, predicts ways in which intelligent machines might cause the extinction of mankind. He says that super-clever machines might argue to themselves that they are superior to humans. They might eventually be put in charge of managing resources and decide that the most efficient course of action is for humans to be removed. He also believes it would be possible for machines to override inbuilt safeguards.

"If you have a very intelligent system it could unprogram itself," he says. "We have to be careful about getting into a situation where they take over against our will or with our blessing."

Even if there exists a distant danger, some experts say it is much too soon to start panicking. Rodney Brooks, director of the Artificial Intelligence Laboratory at Massachusetts Institute of Technology (MIT) says we can't hope to accurately imagine how things may pan out just yet. "I think that this is a little like worrying about noise abatement issues at airports back during mankind's first attempts at a hot air balloon," he says.

Ray Kurzweil, author of *The Age of Spiritual Machines: When Computers Exceed Human Intelligence*, also believes it is possible to overreact to a vision of robotic Armageddon and says the potential benefits make it impossible to turn our backs on the benefits of artificial intelligence.

Handing Control to Machines

If the machines are permitted to make all their own decisions, we can't make any conjectures as to the results, because it is impossible to guess how such machines might behave. We only point out that the fate of the human race would be at the mercy of the machines. It might be argued that the human race would never be foolish enough to hand over all the power to the machines. But we are suggesting neither that the human race would voluntarily turn power over to the machines nor that the machines would willfully seize power. What we do suggest is that the human race might easily permit itself to drift into a position of such dependence on the machines that it would have no practical choice but to accept all of the machines' decisions. As society and the problems that face it become more and more complex and machines become more and more intelligent, people will let machines make more of their decisions for them, simply because machine-made decisions will bring better results than man-made ones. Eventually a stage may be reached at which the decisions necessary to keep the system running will be so complex that human beings will be incapable of making them intelligently. At that stage the machines will be in effective control. People won't be able to just turn the machines off, because they will be so dependent on them that turning them off would amount to suicide.

Theodore Kaczynski, "The Unabomber's Manifesto," n.d.

"People often go through three stages in examining the impact of future technology," says Kurzweil in an article responding to Bill Joy's polemic, titled *Promise and Peril: Deeply Intertwined Poles of Twenty First Century Technology*. "Awe and wonderment at its potential to overcome age old problems, then a sense of dread at a new set of grave dangers that accompany these new technologies. Followed, finally and hopefully, by the realisation that the only viable and responsible path is to set a careful course that can realise the promise while managing the peril."

Welcoming Intelligent Machines

Surprisingly, there are even experts who would welcome the possibility of machines taking over from humans. Professor Hans Moravec is well known for his belief that machines will inherit the earth—he even welcomes the prospect. Moravec said in a recent interview that the majority of significant human evolution has taken place on a cultural level and therefore replacing biological humans with mechanical machines capable of far greater learning and cultural development is the next logical step in evolution.

So what may be the best course of action? Marvin Minsky is an artificial intelligence pioneer who founded the Artificial Intelligence Lab at MIT and is on the board of advisors at the Foresight Institute, a body created to investigate the dangers of emerging technologies. Minsky agrees that extinction at the mechanical hands of a robot race may be just around the corner, but says that developments in the field of artificial intelligence call for considered debate. He says he is encouraging artificial intelligence experts to participate in the work of the Institute.

"Our possible futures include glorious prospects and dreadful disasters," says Minsky in an email. "Some of these are imminent, and others, of course, lie much further off."

Minsky notes that there are more immediate threats to think about and combat, such as global warming, ocean pollution, war and world overpopulation. However, he says, the possibilities of artificial intelligence should not be completely ignored.

"In a nutshell, I argue that humans today do not appear to be competent to solve many problems that we're starting to face. So, one solution is to make ourselves smarter—perhaps by changing into machines. And of course there are dangers in doing this, just as there are in most other fields—but these must be weighed against the dangers of not doing anything at all."

Minsky adds a warning for those who question whether machines may ever become intelligent enough to better us. "As for those who have the hubris to say that we'll 'never' understand intelligence well enough to create or improve it, well, most everyone said the same things about 'life'—until only a half dozen decades ago."

> "*We aren't used to contraptions being intelligent, and are willing to credit them with possession of the full orchestra of creaturehood on hearing a few flute-like notes.*"

Technology Will Not Threaten Society

Chris Malcolm

Recent advances in robotics and artificial intelligence have led some to perceive an imminent threat to the human race. Some allege that computers may surpass humans in intelligence and independence, which may lead to a loss of human control over the planet. In the following viewpoint, Chris Malcolm maintains that the creation of artificial intelligence poses no threat to humans. He contends that while computers will become much more powerful, they will not necessarily become more intelligent, and they will not possess an instinct for survival. Malcolm is a lecturer in the School of Artificial Intelligence of the Division of Informatics of Edinburgh University.

As you read, consider the following questions:
1. As quoted by Malcolm, how does professor de Garis perceive the future of robotics?
2. What are the two extremely difficult areas of artificial intelligence research, according to the author?
3. What does the author claim is the significance of Vaucanson's invention in 1738?

Reprinted, with permission, from "Why Robots Won't Rule," by Chris Malcolm, a Web article published December 10, 2000, at www.dai.ed.ac.uk/homes/cam/Robots_Wont_Rule.shtml.

There is a currently popular argument that within a few to several decades robots (or some other kind of intelligent machine) will have become so much more intelligent than us that they will take over the world. This argument is seriously put forward by knowledgeable scientists working in appropriate disciplines. They take different attitudes to this future. For example, Professor Moravec (a roboticist from Carnegie Mellon University, US) thinks this will be good, because we will be handing the torch of future civilisation over to our "children". Professor Warwick (a roboticist from Reading University, UK) thinks they may snatch the world from us before we are willing to hand it over. Professor de Garis (head of the Artificial Brain Project of Starlab, Belgium) thinks there will be a war between those who are on the side of the robots and those who are against them. Kurzweil (developer of some of the world's most advanced speech synthesisers and recognisers) thinks that we can participate in this superintelligence by having microscopic nanocomputers link themselves into our brains. What they all do agree about is the inevitability of some kind of superintelligent machine soon becoming vastly more intelligent than us.

Like all the best conjuring tricks, the argument depends on distracting you with astonishing facts while some assumptions sneak past.

Escalating Technology

The astonishing facts are a generalisation of Moore's Law. In 1965 Gordon Moore, then of Fairchild (later to found Intel) predicted that the amount of transistors packable into a chip would double every year. It turned out to be closer to 18 months. It affects both computer processors and their associated on-board memory, i.e., the two components most responsible for what we think of as "computer power". For various architectural reasons this permits the computer power of PCs to double every 18 months.

Computer scientists predict that the silicon chip technology on which current computers are based has another ten to twenty years left before it hits fundamental physical limits beyond which no further progress in miniaturisation will be possible. What then? In fact, as Moravec has shown, Moore's Law can be projected backwards since before the dawn of "silicon chips", right back to clockwork. Moravec also normalises the data to "processing power per $1000 (1997)" to produce a "bang per buck" version of Moore's Law. When the data is plotted it can be seen that Moore's Law has leapt seamlessly from technology to technology, always finding a new one before the old one ran out of steam. This suggests that Moore's Law is a specific example of some deeper law concerning information processing technologies in general. So, if this trend persists, we can expect Moore's Law to keep going, leaping technologies again, and again, and again.

Forever? It turns out that we needn't worry about forever, because something very interesting indeed happens in the next few decades. Within a few decades $1000 (1997) will be able to buy a computer with the processing power of the human brain, according to our current best estimates of what that is. Such is the magic of this kind of exponential growth of computer power (doubling every 18 months) that it doesn't matter if we have underestimated the power of the human brain by a factor of 100. We only have to wait another ten years for these $1,000 computers to be 100 times more powerful. Would you prefer to wait until the computers were as powerful as the summed brain power of the entire planetary human population of six billion people? You will just have to wait another 50 years.

In short, we have somehow managed to get ourselves onto a technological escalator which will produce cheap computers of superhuman processing power within a few to several decades. This is the astonishing fact: computers are soon likely to outstrip the processing power of the human brain.

Common Sense and Rapid Learning

The first assumption which sneaked past is that this increase in computer processing *power* will automatically mean an increase in the *intelligence* of whatever is using these computers for brains. As Moravec has shown, this is what has happened so far in a number of areas in robotics and artificial intelligence. In order for machine intelligence to keep step with machine computer power, however, we need a much stronger argument than that it often happens. We need to be able to say that it *always* will, that this is a general rule to which there are no exceptions.

Robots Will Not Surpass Humans

While some of the intelligence of robots is associated with the machine, other aspects of robot and computer, unlike humans', can move quickly and easily from place to place. Some aspects can even combine with data or behaviors from other "agents." Thus, highly intelligent robots probably will have less ego identity associated with a particular body or "set membership" than do human beings. Eventually, human intelligence may also escape its ensnarement in biological tissue and be able to move freely across boundaries. To move from aspects of human intelligence would require virtual simulations of various body parts, but we may find that some aspects of human intelligence are also corpus independent. In some yet far distant future combining human and machine intellect, we may find the concept of discrete beings difficult to understand—or it may only apply at a metalevel—but back to today and the reality of our lifetimes.

Jeanne Dietsch, *ActivMediaRobotics*, 2000.

Unfortunately there are exceptions. Artificial Intelligence has achieved many successes with the kind of canned "intelligence" exemplified by Expert Systems which capture the expertise of human expert, often a consultant diagnostician such as a medical specialist. These are, however, notoriously fragile, falling apart quite idiotically when moved slightly beyond their area of expertise. In other words, they lack the general underpinning of "common sense" which we have. They are also incapable of the widely general, insightful, and rapid learning which characterises human students.

These two areas, of common sense and machine learning, are generally recognised in artificial intelligence to be extremely difficult research areas of which we are only just beginning to scratch the surface. They are also generally recognised to be crucial to the development of human-scale intelligence. Our progress in these areas at the moment consists largely of finding out how much more complex they are than we first supposed.

Optimists Minimize Difficulties

The most important single lesson which artificial intelligence has learned in its 50 years of research is a generalisation of Hofstadter's Law of software development: the problem is much more difficult than you think, even when you take this into account. In other words, the optimists do not have a good track record.

That is why machine intelligence will not follow the development of computational power.

Even if it did, however, that is still not enough to permit the "robots will take over" scenario, because the second assumption which sneaked past is that something which displays some of the attributes of creaturehood must be a real creature. We are strongly disposed by evolution and by habit to suppose that anything displaying some aspects of animate behaviour is animate. It's a usefully cautious assumption in a dangerous world. The point about creatures is that millions of years of evolution have equipped them with a fierce determination to survive. This involves such things as attacking other creatures who threaten their territory.

Intelligence is no more enough to make a creature than is fur and beady eyes. No matter how much intelligence is added to your word processor it is not going refuse to edit any more letters if you don't improve your spelling. And no matter how much intelligence you add to your washing machine, robot butler, or whatever, it is not going to become anything more than a smarter contraption. Our problem is that while we have got used to the idea that teddy bears are not real, we aren't used to contraptions being intelligent, and are willing to credit them with possession of the full orchestra of creaturehood on hearing a few flute-like notes.

If It Looks Like a Duck . . .

This is like what happened with Vaucanson's famous mechanical duck, the duck which aroused such controversy that it still features today in the saying "if it walks like a duck, and quacks like a duck, then it is a duck." In 1738 Vaucanson exhibited his marvelous mechanical duck to an astonished Paris. It had multiply jointed realistic wings, could move its head around and mimic the swallowing neck movements of a duck, "eat" grain, splash water, etc. The Parisians were used to ingenious clockwork automata which played whistles, wrote with pen on paper, etc., but what astounded them about this duck, and convinced them that it was a real step forward towards artificial life was that it had guts made of rubber hose and actually shat evil smelling duck turds soon after eating. Although it was fastened to a large plinth full of the gears and pulleys that made it work, the press of the day, being just as gullible as today's concerning these matters, soon had it capable of walking and swimming and nourishing itself on grains.

In 1976 Joseph Weizenbaum, author of one of the early attempts at passing the Turing Test for Artificial Intelligence, the well-known "Eliza" conversational program (some of its incarnations known as "Doctor" because it emulated the sympathetic enquiries of a psychotherapist), decided that he had seen so much human gullibility and anthropomorphisation towards what was just a bag of barely plausible text manipulation tricks that he concluded that the human race was simply not intellectually mature enough to meddle with such a seductive science as artificial intelligence. We would simply make dreadful fools of ourselves by anthropomorphising and over-interpreting everything.

I'm afraid that Messrs de Garis, Kurzweil, Moravec, Warwick, etc., have proved Weizenbaum all too prescient. I presume of course that the fact that publishers and TV programme makers want to hear about robots taking over, and don't want to hear about robots not taking over, has nothing to do with it.

It takes more than quacking and shitting to make a duck.

"*We are moving closer to [George] Orwell's nightmare: the truth ceaselessly modified, altered, edited, or altogether obliterated.*"

Technology May Affect Human Nature

Joseph D'Agnese

In the following viewpoint, Joseph D'Agnese claims that future technology will bring drastic changes to society and to the individual. He contends that genetic engineering, robotics, and nanotechnology—building electronic circuits and devices from single atoms and molecules—threaten humans by turning richly diverse personalities into automated machines. D'Agnese is a contributing editor at *Discover* magazine, a science and technology publication for the general public.

As you read, consider the following questions:

1. How does the author describe driving a car in the year 2020?
2. What does D'Agnese mean when he says that technology "favors some and eclipses others"?
3. How does the Web resemble the human unconscious, according to the author?

Excerpted from "What You'll Need to Know in Twenty Years That You Don't Know Now," by Joseph D'Agnese, *Discover*, October 2000. Reprinted with permission.

You know things, you child of the 21st century. You may not stop to think about it, but you know stuff to get along that your hallowed progenitors could never have dreamed of. You know how to delete. You can pull down a menu. You know how to change the channel on your TV without getting out of your overstuffed chair, a luxury your grandparents did not enjoy. You know—or should know—how to keep your kid from downloading porn on the Internet. You even know the word download. You know other words too. Scary words. Ebola, mad cow, West Nile virus. At the very mention of these words, your mind knows to give you the creeps.

As always, knowledge is power. What you know gets you through your day. Protects your family. Keeps them safe.

And what you need to know has changed—a lot. Twenty years ago, you took notes with a pen, not a pointer pushed across the face of a personal digital assistant. Twenty years ago you still thought a mouse was just a rodent. Had some prescient soul sidled up to you on the street in 1980 and said, "Listen, buddy, soon you're going to need to know how to operate a big glowing box on your desk by sliding a plastic thing around," you would have seen him as a madman, not a prophet.

By the year 2020, you will need to know stuff you can hardly guess today. You've heard all this before. You're hip to this technology thing. You know the drill: Gadgets change; you adapt to the gadgets. You learn new buzzwords, talk the talk, and keep on going as humans have for millennia. But in doing so, you reduce technology to a heap of glorified hand tools, the equivalent of sticks chimps use to extract ants from a hole.

A Technological Circus

In 20 years it should be painfully clear that technology never just hands us tools; it grants us a passport to a world where choices multiply, desires are ignited, and new moral decisions confront us. Like an odyssey-starved traveler, you'll wander through a kind of exotic street fair, senses assaulted by too much information. You'll find yourself frazzled just trying to draw a bead on all the options, while a numbing calliope tune insists it's all great fun. A lot of what you'll encounter will just be cool, requiring nothing more than nimble minds and fingers. By the year 2020, for example, you will need to know how to talk to your house. Today your home contains dozens of appliances, each working independently. But someday you'll cross the threshold and everything will know you're home. The lights will flicker on, the air conditioner will have kicked in, the refrigerator will clamor to enumerate all the meals you can assemble with the groceries cached inside. In exchange for this convenience, you'll share your abode with a horde of circuit-heavy infants that constantly burble to each other and cry out for your care. If you're the one with mechanical flair at your house, when you come through the door at night, your spouse will be pulling you aside to whisper, "Honey, I need you to talk to the robot."

By the year 2020, you will have to learn to drive a more automated car. You'll get behind the wheel of a smart car that avoids fender benders by braking before you even see danger looming. At a much later date, you will slip into a bucket seat as if at the movies—snacks, reading material, and sodas at the ready—sit back, relax, program the car, and over the freeways to grandmother's house you'll go. Like the toaster and coffeemaker back home, the car's sensors will monitor the activity and destinations of other cars on the road. "Going my way?" your vehicle will bleep in autospeak. "Indeed," responds the living room on wheels in the left lane. And the two will hitch up and rocket toward their common goal together. This technology will conserve fuel and may save lives, but the pleasure of driving as you know it will be gone. That's something you should know.

But perhaps by now you've realized that for every convenience technology bestows upon us, it chips away at something else. All of us, great souls as well as lost ones, must in time wrestle with this notion. If you are the poet William Blake, in metered rhyme you decry the Satanic mills; if you are Ted Kaczynski [the Unabomber], you take to the hills and spit death by snail mail. Most of us simply acknowledge the trade-offs and move on.

Each time we do this, though, we march farther away from

a world we can touch and comprehend in our bones toward one that we pray will work better. Consider: In the year 2020, you'll identify yourself, gain access to homes and businesses, and board aircraft after a laser has measured the shape of your irises. But the price will be loss of privacy. A record of your transactions, your daily comings and goings, will be just a keyboard tap away from others.

Loss of Privacy

Booting up your home PC has already become a public act. Meander the Web today, and almost every move you make is cataloged in service to the gods of commerce. They know what you're buying. What you listen to. Where you chat. By 2020 you'll need to know how to dean up that electronic trail day in and day out. "Say you were searching for information on hats" theorizes Jaron Lanier, computer scientist, musical composer, and virtual reality pioneer, "and you saw a link about hats, but when you got to it, it was actually a weird pornography site about hat fetishes. Then it turns out there's a record that you visited this site, and now you're getting bombarded with offers from people with hat fetishes. Furthermore, your friends are being contacted in case they have hat fetishes. All of a sudden you're the hat fetish person in your social circle, and you have to go in and undo it." To throw the hounds off your scent, Lanier says, you could spend the afternoon downloading the Great Books or posing as a do-gooder in search of charities deserving of your drachmas.

In time, you'll be wielding electronica for the same reasons medieval crusaders took up sword and lance: to ward off intruders. Rooting out destructive viruses and spam in your equipment will become old hat, as will the regular checks you'll be performing on your groceries and yourself. Tomorrow's Kaczynskis will be able to concoct harmful viruses and insinuate them into the food supply, or perhaps release pathogens in public places. You'll need to be ready for them. Daily computer checkups of your blood, saliva, or bodily waste will be effortless, the medical equivalent of checking your stock portfolio. "Real-time monitoring," says James Weiland, assistant professor of ophthalmology at Johns Hopkins, "will tell you in the morning what vitamin your body is low on and what to have for breakfast."

With all this new information, you'll stand a better chance of living well beyond your biblical allotment of threescore and ten. More than 200,000 centenarians will inhabit the United States in 2020—why shouldn't you be one of them? To reach that age you'll need to know enough to make more complicated medical choices: Do I want to jettison a limb and wait five years to regrow another? Shall I allow a phalanx of nanobots to scrape the plaque out of my arteries or opt to replace the vessels altogether? "Amateurs may be fooling around with black-market genetic manipulation," says Marvin Minsky, one of the founders of the Artificial Intelligence Lab at MIT, "maybe extending their lives by lengthening their own telomeres, the ends of chromosomes believed to control life span. Or they might, in fact, be growing new features in their brain."

Technology and Its Consequences

With technology in general, we have to begin to ask what it is, what are its consequences? I think we have to begin to no longer think of it as neutral, and we need to begin to understand that it embodies its consequences, both good and bad. And information technologies have an incredibly wonderful power to them, but they do also deliver violence steadily to children, so it also has its darker side. Genetic technologies are truly the most revolutionary of the technologies that are coming along. It is the thing that is going to shape the next century. And the most fundamental question for us in the next 100 years is really, what does it mean to be human? We need to put these technologies in the human context to understand how they can best be applied.

Nana Naisbitt, "Technology 2000," Newshour with Jim Lehrer Transcript, December 29, 1999.

By the year 2020, science will understand the Creator's software well enough to tell you a great deal about the genetic hand dealt you and those you love. Science may even help you decide if you should quit loving them. These days it's not unheard of for one partner to investigate the other's background or assets before marrying. In the future you'll need to access your betrothed's genetic map, see what diseases he or she is likely to contract, assess the appearance and health of your children, and perhaps even size up your love's mental health. Of course, this swings both ways. In this world, you will be forced to ask: Do I want to know if I'm earmarked for heart disease or breast cancer? Do I want my potential spouse to know? If I know this, and my doctor knows, does it mean that my insurance carrier must know? If this last one scares you, it should. It could mean the end of health care as you know it.

Designer Genetics

This is just the beginning. Once we know the future, we're going to be tempted to rewrite the software. Clearly, it would be an act of kindness to reach into that fragile, permeable, four- or eight-cell being and rid it of the disease that cut short the life of its great-grandfather. But why wait for conception? Why not design your kid, toes up, out of whole cloth: the blue-eye gene, the blond-hair gene, the excel-at-lacrosse gene. Ban such tinkering, and citizens will merely scurry underground in order to conceive the perfect child.

If we can tear ourselves away from such selfish goals long enough to look around, we will have to face the fact that technology favors some and eclipses others. Bill Robinson, who spent 30 years as an electrical engineer with Canada's Nortel Networks, has been thinking about this issue recently. "We spend our time and effort creating exciting new communications technologies," he writes, "yet half the world does not have access to a telephone. We use the Internet to order the latest novel, yet many people in the world don't have access to books. We are now discussing embedded processors to connect our refrigerators to bathroom scales and the grocery store, yet many children in the world go to bed hungry at night."

This grisly reality will be harder to hide from when our planet swells to 8 billion people in 2020. For Lanier, the most heartbreaking scenario is festering in the third world, where, he believes, the current generation of children—lacking food, lacking skills, lacking aid, lacking education—will be lost in the next techno-revolution. "What is going to happen to all these people as they start to age, say, 20 years from now?" he wonders. "You're going to have to somehow live while you watch a billion people starve, which is going to be a new human experience. How will we do that?"

Good question. And just one of many difficult questions waiting. How can I choose between two genetic scripts for a child I have yet to know? How much of myself should I reveal on the Web? How will I cope with all these machines when they break down, including the self-replicating nanopests that may be residing in my flesh? In our zeal to be happy little technologists, we'll turn, much as we do today, to the Web for answers. And we'll perfect the art of being disappointed.

A Science Fiction Nightmare

If any medium ever resembled the human unconscious, the Web is it: a place of hidden wonders, stray inane thoughts, peaks of brilliance, valleys of perversity. And no apparent governor. Type your query, hit return, and voila!—10,000 hits. Good luck shaking them down. Even in 2020 you will always need to know if the facts you've dredged up are accurate and truthful. With so many sources doling out information, you will need to know: What is he selling, and why is he selling it? Most unsettling is the fact that these precious touchstones are not permanent. They never will find their way to the library stacks. Instead we are moving closer to George Orwell's nightmare: the truth ceaselessly modified, altered, edited, or altogether obliterated. Here today, gone tomorrow, with nothing but a bewildering ERROR 404 FILE NOT FOUND left in its place.

By then, you will no longer be a child of the 21st century. If anything, you'll be an elder, your mind and body augmented, your chromosomes refreshed, flexible computers woven into the four corners of your garments. On the one hand, your workload will multiply as you bat away each glitch resulting from the increased number of gadgets in your life. On the other, you will be forced to take on moral questions no human has ever faced. When will you find time to do that? How will you contemplate when everything is speeding up and time for reflection is practically nonexistent?

That's you in 20 years. Like the machine that inspired your age, you will be constantly scanning, processing, sifting, searching for a code to guide you through. And yet the key, the compass, the answer, was once offered in a temple at Delphi. What will you need to know in 2020? Yourself.

> *"The future . . . will not be unrecognizably
> exotic because across all the dizzying
> changes that shaped the present and will
> shape the future one element remains
> constant: human nature."*

Technology Will Not Affect Human Nature

Steven Pinker

Futurists debate how the human psyche will respond to the numerous technological conveniences just beginning to permeate everyday life. Many argue that the attitudes and behaviors of the future will be unrecognizable compared to contemporary society. In the following viewpoint, psychology professor Steven Pinker contends that human nature will remain essentially intact and that the age-old questions of philosophy and religion will continue to enchant and mystify future generations. Because human nature remains constant, according to Pinker, the future will not be as foreign as some believe it will be.

As you read, consider the following questions:
1. Why does the author contend that the human biological constitution will not change?
2. According to Pinker, why will humans retain control over computers?
3. What are three predictions of futurologists that frighten people away from technology, according to the author?

Reprinted from "Life in the Fourth Millennium," by Steven Pinker, *Technology Review*, May/June 2000, by permission of *Technology Review* via the Copyright Clearance Center.

People living at the start of the third millennium enjoy a world that would have been inconceivable to our ancestors living in the 100 millennia that our species has existed. Ignorance and myth have given way to an extraordinarily detailed understanding of life, matter and the universe. Slavery, despotism, blood feuds and patriarchy have vanished from vast expanses of the planet, driven out by unprecedented concepts of universal human rights and the rule of law. Technology has shrunk the globe and stretched our lives and our minds.

How far can this revolution in the human condition go? Will the world of 3000 be as unthinkable to us today as the world of 2000 would have been to our forebears a millennium ago? Will our descendants live in a wired Age of Aquarius? Will science explain the universe down to the last quark, extinguishing mystery and wonder? Will the Internet turn us into isolates who interact only in virtual reality, doing away with couples, families, communities, cities? Will electronic media transform the arts beyond recognition? Will they transform our minds?

Obviously it would be foolish to predict what life will be like in a thousand years. We laugh at the Victorian experts who predicted that radio and flying machines were impossible. But it is just as foolish to predict that the future will be utterly foreign—we also laugh at the postwar experts who foresaw domed cities, jet-pack commuters and nuclear vacuum cleaners. The future, I suggest, will not be unrecognizably exotic because across all the dizzying changes that shaped the present and will shape the future one element remains constant: human nature.

The Brain as a Toolbox

After decades of viewing the mind as a blank slate upon which the environment writes, cognitive neuroscientists, behavioral geneticists and evolutionary psychologists are discovering instead a richly structured human psyche. Of course, humans are ravenous learners, but learning is possible only in a brain equipped with circuits that learn in intelligent ways and with emotions that motivate it to learn in useful ways. The mind has a toolbox of concepts for space (millimeters to kilometers), time (tenths of seconds to years), small numbers, billiard-ball causation, living things and other minds. It is powered by emotions about things—curiosity, fear, disgust, beauty—and about people—love, guilt, anger, sympathy, pride, lust. It has instincts to communicate by language, gesture and facial expressions.

We inherited this standard equipment from our evolutionary ancestors, and, I suspect, we will bequeath it to our descendants in the millennia to come. We won't evolve into bulbous-brained, spindly-bodied homunculi because biological evolution is not a force that pushes us to greater intelligence and wisdom; it simply favors variants that outreproduce their rivals in some environments. Unless people with a particular trait have more babies worldwide for thousands of generations, our biological constitution will not radically change.

It is also far from certain that we will redesign human nature through genetic engineering. People are repulsed by genetically modified soybeans, let alone babies, and the risks and reservations surrounding germ-line engineering of the human brain may consign it to the fate of the nuclear-powered vacuum cleaner.

If human nature does not change, our lives in the new millennium may be more familiar than the futurologists predict. Take education, where many seers predict a revolution that will make the schoolroom obsolete. Some envision . . . free schools, where children interact in a technology-enriched environment and literacy and knowledge will just blossom, free from the drudgery of drill and practice. Others hope that early stimulation, such as playing Mozart piano concertos to the bellies of pregnant women, will transform a plastic brain into a superlearner.

Stretching Stone-Age Minds

But an alternative view is that education is the attempt to get minds to do things they are badly designed for. Though children instinctively speak, see, move and use common sense, their minds may be constitutionally ill at ease with many of the fruits of modern civilization: written language, mathematical calculation, the very large and very small spans of time and space that are the subject of history and science. If so, education will always be a tough slog, depending on disciplined work on the part of students and on the insight of a skilled teacher who can stretch stone-age minds to meet the demands of alien subject matter.

Our mental apparatus may also constrain how much we adults *ever* grasp the truths of science. The Big Bang, curved 4-D space-time and particles that act like waves—all are required by our best theories of physics but are incompatible with common sense. Similarly, consciousness and decision-making arise from the electrochemical activity of neural networks in the brain. But how moving molecules should throw off subjective feelings (as opposed to mere intelligent computations) and choices for which we can be held responsible (as opposed to behavior that is caused) remain deep mysteries to our Pleistocene psyches.

That suggests that our descendants will endlessly ponder the age-old topics of religion and philosophy, which ultimately hinge on concepts of matter and mind. Why does the universe exist, and what brought it into being? What are the rights and responsibilities of living things with different brains, hence different minds, from ours—fetuses, animals, neurologically impaired people, the dying? Abortion, animal rights, the insanity defense and euthanasia will continue to agonize the thoughtful (or be settled by dogma among the unthoughtful) for as long as the human mind confronts them.

Shaping Information Technology

One can also predict that the mind will shape, rather than be reshaped by, the information technology of the future. Why have computers recently infiltrated our lives? Because they have been painstakingly crafted to mesh better with the primitive workings of our minds. The graphical user interface (windows, icons, buttons, sliders, mice) and the World Wide Web represent the coercion of machines, not people.

We have jiggered our computers to simulate a world of phantom objects that are alien to the computer's own internal workings (ones, zeroes and logic) but are comfortable for us tool-using, vision-dependent primates. Many other dramatic technological changes will come from getting our machines to adapt to our quirks—understanding our speech, recognizing our faces, carrying out our desires in accord with our common sense—rather than from getting humans to adapt to the ways of machines.

Our emotional repertoire, too, ensures that the world of tomorrow will be a familiar place. Humans are a social species, with intense longings for friends, communities, family and spouses, consummated by face-to-face contact.

Innovation Comes from Human Imagination

Today's technological revolution is a cycle of innovation, moving at incredible speed to reshape the way we work, learn, play, and communicate. As we seek ways to manage that change, we should first remember that technology itself has no power to drive the cycle. Innovation comes not from the inventions, but from human imagination, creativity and action.

Bernard Verghes, *OECD Observer*, Summer 2000.

E-mail and e-commerce will continue their inroads, of course, but not to the point of making us permanent antisocial shut-ins; only to the point where the increase in convenience is outweighed by a decrease in the pleasure of being with friends, relations and interesting strangers. If our descendants have spaceports and transporter rooms, they will be crammed at Thanksgiving and Christmas.

Human Conflicts Persist

But human relationships also embrace conflicts of biological interests, which surface in jealousy, sibling rivalry, status-seeking, infidelity and mistrust. The social world is a chess game in which our minds evolved as strategists.

If so, the mental lives of our descendants are not hard to predict. Conflicts with other people, including those they care the most about, will crowd their waking thoughts, keep them up at night, animate their conversation and supply the plots of their fiction, whatever the medium in which they enjoy it.

If constraints on human nature make the future more like the present and past than futurologists predict, should we sink into despair? Many people, seeing the tragedies and frustrations of the world today, dream of a future without limits, in which our descendants are infinitely good, wise, powerful and omniscient. The suggestion that our future might be constrained by DNA shaped in the savanna and ice ages seems depressing—even dangerous.

Admittedly, many declarations of ineluctable human nature turned out to be wrong and even harmful—for example, the "inevitability" of war, racial segregation and the political inequality of women. But the opposite view, of an infinitely plastic and perfectible mind, has led to horrors of its own: the Soviet "new man," reeducation camps and the unjust blaming of mothers for the disabilities and neuroses of their children.

Recognizing Human Needs

Many leaps in our quality of life came from the recognition of universal human needs, such as life, liberty and the pursuit of happiness, and of universal limitations on human wisdom and beneficence, which led to our government of laws and not men.

Universal obsessions are also the reason that we enjoy the art and stories of peoples who lived in centuries and millennia past: Shakespeare, the Bible, the love stories and hero myths of countless cultures superficially unlike our own. And the mind's foibles ensure that science will be a perennial source of enchantment even as it dispels one mystery after another. The delights of science—of the Big Bang, the theory of evolution, the unraveling of the genes and the brain—come from the surprise triggered by a conclusion that is indubitably confirmed by experiment and theory but that contradicts standard human intuitions.

Third-millennium futurologists should realize that their fantasies are scaring people to death. The preposterous world in which we interact only in cyberspace, choose the endings of our novels, merge with our computers and design our children from a catalogue gives people the creeps and turns them off to the genuine promise of technological progress. The constancy of human nature is our reassurance that the world we leave to our descendants will be one in which scientific progress leads to delight rather than boredom, in which our best art and literature continues to be appreciated, and in which technology will enrich rather than dominate human lives.

> "Our leaders should begin to restructure
> social, economic and government systems to
> accommodate the forthcoming change."

Society Must Be Restructured to Accommodate Technology

William H. Davidow

In the following viewpoint, William H. Davidow argues that information technology threatens to undermine the power of government in relation to special interest groups, corporations, and taxpayers. He contends that society must be legally, economically, and socially restructured in order to address this power imbalance. Davidow, a former senior vice president of Intel, is a general partner in Mohr, Davidow Ventures, a Silicon Valley venture capital firm.

As you read, consider the following questions:
1. How does the author differentiate between physical and intellectual infrastructures?
2. What solution does Davidow offer to combat corporate "governments" from gathering too much power?
3. What taxes are the easiest to collect, according to the author?

Reprinted, by permission of the author, from "How Technology Is Turning Society Upside Down," by William H. Davidow, *The Washington Post*, June 6, 1999, p. B-1.

P eople make tools, and tools then remake social systems—
that has been the case for millenniums. The heavy plow
fostered the manorial system of governance in northern Europe in the 9th century, allowing for the accumulation of surplus, population growth and urbanization. The automobile gave us freedom to travel and commute, and in the process hollowed out cities and created suburbs.

How will computers remake today's social systems? No one can say for sure. Certain major elements of change—from how schoolchildren do research for papers to how and where we spend our working hours—are already apparent, of course. But the important issue facing us is this: Do we accept the consequences passively, or try to anticipate and control them? I would go so far as to suggest that our leaders should begin to restructure social, economic and government systems to accommodate the forthcoming change.

For all of human history, society has been built predominantly on a physical foundation of tangible objects and assets. Governments drew their strength from geography: Mountains, rivers and oceans protected them and defined their reach. Businesses depended on their physical assets—factories, inventory, natural resources. Communities were clustered around churches, schools and places of public business. Factories provided places to work. Streets supported the movement of goods.

Physical infrastructure serves a second important purpose: It acts as an information storage and transmission network. Office buildings facilitate communication between co-workers. File cabinets store information. Streets carry information in the form of people, letters, newspapers.

From a Physical to an Intellectual Infrastructure

But for the past several centuries, the importance of society's physical foundations has been gradually eroding as intangible assets such as knowledge, organizational structures, methods of production, intellectual property and trade secrets have become more and more essential. As people developed more accurate and sophisticated methods of navigation and gained knowledge of world markets, countries were no longer dependent on raw materials located entirely within their borders. Improved methods of communication by telegraph, telephone and over data networks have practically eliminated the need for manufacturers to be located near cities.

As a consequence, cities are no longer needed as centers of production. Today, information technology is turning the gradual displacement of physical assets by intangible ones into a rout. NASDAQ, the New York Stock Exchange's most important competitor, does not even have a trading floor. It exists only electronically—and thus, in a sense, exists nowhere at all.

Reducing the cost of information transactions, and severing the bonds between information itself and the physical means of transporting and storing it, has affected almost every aspect of civilization. Gradually, the influence of government will be further circumscribed and the influence of corporations will expand proportionately. Unchecked by traditional forces, large virtual communities—those that unite their members electronically—will develop significant power over the direction of public policy. Tax systems that depended on physical constraints will wither.

Special Interest Groups

Local "special interest" groups—bound by common religion, ethnicity or political goals—are a fundamental part of democratic society. They add diversity, foster bonds, preserve cultural and religious values and protect the interests of their members. When they act to further their own interests at the expense of others, they are subject to local forces that cause them to moderate their stances: resistant neighbors, editorials in the newspaper, opposition at the polls.

But virtual communities—free of many of these local constraints and founded not so much on the basis of religion or ethnicity as on other narrowly focused interests—are expanding rapidly, and information technology will amplify their power dramatically. A national or international special interest group is very different from a local group supporting the exact same causes. An organized virtual community that collects only a few dollars in dues from each of its millions of members is a very potent force, and is substantially less sensitive to the interests and feelings of others.

It took years to build up the National Rifle Association and the American Association of Retired Persons by means of pre–Information Age procedures. These organizations were subject to the constraints of the physical world—sending actual mail, holding actual conferences, maintaining local chapters in real towns and cities. Freed from such constraints, special interest groups can be expected to achieve critical mass sooner and become larger and more reactive.

One way to confront this problem would be to return more political authority to smaller political jurisdictions. For example, the federal government could let the states control all agricultural subsidies. This would make it more difficult for a virtual community of tobacco farmers to get northern states, for example, to subsidize tobacco production. The federal government should also consider getting out of the education business entirely, and letting states and local school districts set the education agenda. This would in all likelihood curb the ability of virtual political and religious organizations to influence the educational agenda on a nationwide basis.

Corporate Power

The evolving information technology of the past century has given multinational companies the power to control and coordinate activities around the world. It has enabled the matching of the production of goods around the globe to worldwide demand, and facilitated the design of products in one location and their production in another.

Even when "information technology" consisted mainly of the telephone, multinational corporations became more difficult for governments to control. The companies could structure their operations to minimize taxes by moving profits to geographic locales that provided favorable treatment. When regulations increased costs, they could easily move production to other locations. But the advent of computers has enhanced these capabilities incalculably, and brought them within reach of even modest companies.

Not too long ago, governments thought they could control their national economies. They believed they could increase exports by devaluing their currencies and create jobs through deficit spending. Today, much of that power has been taken over by multinationals. If one considered multinational companies to be economies, 50 of them would be listed in the top 100 of the world's largest economies.

These companies have the ability to decide where the jobs will be and who exports and who imports. They determine the trade balance in the industry segments they control. By choosing the countries in which to locate, they decide, in effect, what laws they will obey. They have the flexibility to employ child labor, to pay very low wages, and to select locations that are willing to trade environmental damage for jobs.

National governments that wish to manage the situation should consider attempting to control the actions of multinational corporate "governments" through agreements rather than just laws, as the ordinary regime of national laws and regulations will become less and less effective.

The Collapse of Taxation

Information technology poses a grave threat to the tax system, which depends not only on an individual's honesty, but also on the constraints imposed on those individuals and corporations by physical infrastructure. One of the fundamental requirements for a government's tax schemes is a knowledge of where money is earned, where a transaction takes place and where value is added. A second requirement is the ability to exert authority over those activities.

Information technology undermines each of these preconditions. Suppose a customer in the United States orders a product from a company located in a tax haven in the Caribbean and uses funds in an offshore bank to pay for the purchase. The company in the Caribbean sends an electronic message to its factory in Mexico to manufacture the product and ship it to the customer. With a little bit of financial engineering, this transaction becomes nearly impossible to tax. It is not clear where the transaction takes place; in fact, it would be extremely difficult for a government to determine if a product had been ordered at all. The type of transaction just described has been possible for some decades, by phone and wire. But as corporations have become more far-flung and communications more sophisticated, the opportunity for such transactions has expanded enormously. While these problems are not entirely new, information technology increases their intensity.

Individuals, too, have new freedom to avoid taxes—both legally and illegally. In the past, many professionals tended to locate near the consumers of their services or near the sources of information critical to their practice. But today, many engineers, financial managers, lawyers and others have the flexibility to live and work in states that have little or no personal income tax while at the same time continuing to serve clients elsewhere.

In general, owing to information technology, income taxes collected on a national level, whether they be on individuals or corporations, will be the most difficult taxes to collect—they'll be lost in the nether world of the intangible society. As it becomes easier to avoid taxation and the risks of doing so decline, more organizations and individuals will likely choose this option.

The easiest taxes to collect will be the ones based on activities of consumers that simply must take place locally: Gasoline must be delivered and sold locally, water and electricity have to go to actual businesses and homes, gas and heating oil have to be delivered to the point of consumption. It would be difficult to avoid taxes on real estate, food, transportation, entertainment, hotel rooms and health care. Indeed, many of the taxes that will be easiest to collect are regressive.

Taxes on most forms of consumption fall most heavily on the poor who spend a disproportionate amount of their income on consumables and energy.

If a government wished to get ahead of this problem, it would begin openly and deliberately shifting the tax base to consumers while designing rebate systems to make such taxes less regressive. Dealing with this problem after corporations and professionals have fled the tax base by both legal and illegal means will make confronting it all the more difficult.

We can allow our civilization to accommodate itself, willynilly, to the evolutionary dictates of information technology, as determined by the expediency of the market, or we can attempt to find policies that will influence it. The winners in a civilization remade around computers will not be those who attempt to contain information technology. The winners will be those who invent the new structures of government, business and society in which technology is embedded.

Periodical Bibliography

The following articles have been selected to supplement the diverse views presented in this chapter. Addresses are provided for periodicals not indexed in the *Readers' Guide to Periodical Literature*, the *Alternative Press Index*, the *Social Sciences Index*, or the *Index to Legal Periodicals and Books*.

Chester Burger	"Sooner than You Think," *Vital Speeches*, September 15, 2000.
Mathew Cabot	"Tomorrow's Treatments," *Natural Science*, March 2001.
Steve Case	"Internet's Reach Will Extend Our Grasp, Improve Our Lives," *USA Today*, June 22, 1999.
Economist	"Behold, the Emerald City," *San Diego Union-Tribune*, November 28, 1999.
Peter Godwin	"The Car That Can't Crash," *New York Times Magazine*, June 11, 2000.
Edward Goldsmith	"The Fight Must Go On," *Ecologist*, July/August 2000.
Eric Haseltine	"Twenty Things That Will Be Obsolete in Twenty Years," *Discover*, October 2000.
Bill Joy	"Why the Future Doesn't Need Us," *Wired*, April 2000.
Ray Kurzweil	"As Machines Become More Like People, Will People Become More Like God?" *Talk*, April 2001.
Michel Marriott	"Toys Today, Servants Tomorrow," *New York Times*, March 22, 2001.
Ryan Mathews, Jim Taylor, and Watts Wacker	"The Perishability of Technology," *FirstMatter*, 1999.
Neil Munro	"We're Wired, but Now What?" *National Journal*, January 16, 1999.
Joshua Muravchik	"Machines Are (Sort of) Predictable. Man Isn't," *Wall Street Journal*, December 29, 1999.
Charles W. Petit	"Shape of Things to Come: Are You Ready for a New Sense of Community—Among Machines?" *U.S. News & World Report*, August 17, 1998.
Royal Van Horn	"Technology—What's Next?" *Phi Delta Kappan*, June 2000.
Earl W. Wilkins	"Tomorrow's Technologies: Gazing at the Horizon," *Graphic Arts Monthly*, January 1999.

For Further Discussion

Chapter 1

1. Tom Mahon argues that technology has detracted from humanity and culture by focusing on machines rather than on spirituality and culture. Do you agree with this contention? Why, or why not?

2. V.V. Raman chronicles the technological achievements of the twentieth century and contends that such advances have improved society. Do you think technology has benefited society? How have recent technological innovations made a difference in your life? Explain your answer.

3. Mickey Revenaugh describes the digital divide as a national crisis that requires an immediate solution. Eric Cohen claims that the have and have-not concept of the digital divide is a natural consequence of capitalism. Do you believe that the digital divide is different from any other societal inequity? Citing from the text, compare the digital divide to social divisions stemming from race, income, and single- and two-parent families.

4. James K. Glassman argues that government regulation of technology and the Internet would interfere with the free market and disrupt the economy. How does William J. Ray respond to this argument? With whose perspective do you most agree? Why?

Chapter 2

1. According to Jennifer Brookes, clinical trials are an invaluable tool for medical research. Neal D. Barnard argues that humans involved in such experiments are subjected to dangerous risks. Do you think that the potential for medical breakthroughs supercedes the risk of experimentation? Why or why not?

2. The *Lancet* argues that paying people for organs would not only increase the number of donated body parts, but also fairly compensate sellers in economic distress. How does William E. Stempsey respond to this argument? Whose solution to the organ shortage do you think is most viable? Explain your answer.

3. Steve Connor claims that the organ shortage could be reduced with the use of animal parts instead of human organs. Alan H. Berger and Gil Lamont maintain that the potential dangers of using animal organs, such as disease and infection, pose too many

risks to society. If you or someone in your family needed an organ transplant, would the risks of xenotransplantation be worth taking? Citing from the text, explain why you would or would not accept an animal body part.

4. Lawrence S.B. Goldstein describes the potential of stem cells for the treatment of such degenerative diseases as Alzheimer's and Parkinson's. John Kass argues that research on fetal stem cells is unethical because it recruits an unwitting participant—an aborted embryo—in a medical experiment. Whose argument do you find most convincing, and why?

Chapter 3

1. Jeffrey Rothfeder argues that the convenience of personal computers and the Internet has made data such as financial and medical records vulnerable to privacy violations and fraud. Do you think that convenience is worth the accompanying loss of privacy? Why, or why not?

2. According to Travis Charbeneau, issues that were once considered extremely private, such as homosexuality, have become less private because society has become more tolerant. Do you agree with this argument? Why, or why not?

3. Laura Pincus Hartman describes the controversy over the monitoring of employees in the workplace. Do you think that employers have the right to monitor Internet and e-mail usage by employees? Explain your answer, citing from the text.

Chapter 4

1. Ian D. Pearson claims that trends in technology can be projected to predict its future progress and impact on society. Roger A. Pielke Jr. and Dan Sarewitz disagree and maintain that technological forecasts are as unreliable as weather predictions. Whose argument do you find most convincing? Do you think that current trends in technology are accurate indicators for the future? Explain your answer.

2. Will Knight argues that the creation of robots may jeopardize the existence of humans. Chris Malcolm contends that the lack of a biological survival instinct in robots precludes a robotic coup. Do you think that humans may create technological wonders that may ultimately lead to our destruction? Do you think that a superior but artificial intelligence can overcome human biology? Why or why not?

3. Steven Pinker maintains that history reveals that human nature has changed little over time and that technological conveniences will not taint basic emotions and instincts. Joseph D'Agnese warns against embracing technology too warmly, as humans may become as automated as their computers. With whose argument do you most agree? Do you think that technology affects human nature or human nature affects technology? Explain.

Organizations to Contact

The editors have compiled the following list of organizations concerned with the issues debated in this book. The descriptions are derived from materials provided by the organizations. All have publications or information available for interested readers. The list was compiled on the date of publication of the present volume; the information provided here may change. Be aware that many organizations may take several weeks or longer to respond to inquiries, so allow as much time as possible.

Biotechnology Industry Organization (BIO)
1625 K St. NW, Suite 1100, Washington, DC 20006
(202) 857-0244 • fax: (202) 857-0237
e-mail: info@bio.org • website: www.bio.org

BIO is composed of companies engaged in industrial biotechnology. It monitors government actions that affect biotechnology and promotes increased public understanding of biotechnology through its educational activities and workshops. Its publications include the bimonthly newsletter *BIO Bulletin*, the periodical *BIO News*, and the book *Biotech for All*.

Center for Bioethics and Human Dignity (CBHD)
2065 Half Day Rd., Bannockburn, IL 60015
(847) 317-8180 • fax: (847) 317-8153
e-mail: cbhd@biccc.org • website: www.bioethix.org

CBHD is an international educational center whose purpose is to bring Christian perspectives to bear on contemporary bioethical challenges facing society. Projects have addressed such topics as genetic technologies, euthanasia, and abortion. It publishes the newsletter *Dignity* and the book *Genetic Ethics: Do the Ends Justify the Genes?*

Center for Civic Networking (CCN)
PO Box 65272, Washington, DC 20037
(202) 362-3831 • fax: (202) 986-2539
e-mail: ccn@civicnet.org • website: http://civic.net/ccn.html

CCN is dedicated to promoting the use of information technology and infrastructure for the public good, particularly for improving access to information and the delivery of government services, broadening citizen participation in government, and stimulating economic and community development. It conducts policy research and analysis and consults with government and nonprofit organizations. The center publishes the weekly *CivicNet Gazette*.

Center for Democracy and Technology (CDT)
1001 G St. NW, Suite 700 E, Washington, DC 20001
(202) 637-9800 • fax: (202) 637-0968
e-mail: info@cdt.org • website: www.cdt.org

CDT's mission is to develop public policy solutions that advance constitutional civil liberties and democratic values in new computer and communications media. Pursuing its mission through policy research, public education, and coalition building, the center works to increase citizens' privacy and the public's control over the use of personal information held by government and other institutions. Its publications include issue briefs, policy papers, and *CDT Policy Posts*, an online, occasional publication that covers issues regarding the civil liberties of those using the information highway.

Center for Media Education (CME)
1511 K St. NW, Suite 518, Washington, DC 20005
(202) 628-2620 • fax: (202) 628-2554
e-mail: cme@access.digex.net • website: www.cme.org

CME is a public interest group concerned with media and telecommunications issues, such as educational television for children, universal access to the information highway, and the development and ownership of information services. Its projects include the Campaign for Kids TV, which seeks to improve children's education; the Future of Media, concerning the information highway; and the Telecommunications Policy Roundtable of monthly meetings of nonprofit organizations. CME publishes the monthly newsletter *InfoActive: Telecommunications Monthly for Nonprofits*.

Computing Research Association (CRA)
1875 Connecticut Ave. NW, Suite 718, Washington, DC 20009
(202) 234-2111 • fax: (202) 667-1066
e-mail: info@cra.org • website: http://cra.org

CRA seeks to strengthen research and education in the computing fields, expand opportunities for women and minorities, and educate the public and policy makers on the importance of computing research. CRA's publications include the bimonthly newsletter *Computing Research News*.

Council for Responsible Genetics

5 Upland Rd., Suite 3, Cambridge, MA 02140
(617) 868-0870 • fax: (617) 864-5164
e-mail: info@fbresearch.org • website: www.fbresearch.org

The council is a national organization of scientists, health professionals, trade unionists, women's health activists, and others who work to ensure that biotechnology is developed safely and in the public's interest. The council publishes the bimonthly newsletter *GeneWatch* and position papers on the Human Genome Project, genetic discrimination, germ-line modifications, and DNA-based identification systems.

Electronic Frontier Foundation (EFF)

PO Box 170190, San Francisco, CA 94117
(415) 668-7171 • fax: (415) 668-7007
e-mail: eff@eff.org • website: www.eff.org

EFF is an organization of students and other individuals that aims to promote a better understanding of telecommunications issues. It fosters awareness of civil liberties issues arising from advancements in computer-based communications media and supports litigation to preserve, protect, and extend First Amendment rights in computing and telecommunications technologies. EFF's publications include *Building the Open Road*, *Crime and Puzzlement*, the quarterly newsletter *EFFector Online*, and online bulletins and publications, including *First Amendment in Cyberspace*.

Electronic Privacy Information Center (EPIC)

666 Pennsylvania Ave. SE, Suite 301, Washington, DC 20003
(202) 544-9240 • fax: (202) 547-5482
e-mail: info@epic.org • website: www.epic.org

EPIC advocates a public right to electronic privacy. It sponsors educational and research programs, compiles statistics, and conducts litigation. Its publications include the biweekly electronic newsletter *EPIC Alert* and various online reports.

Human Genome Project (HGP)

Human Genome Management Information System
Oak Ridge National Laboratory
1060 Commerce Park MS 6480, Oak Ridge, TN 37830
(805) 576-6669 • fax: (423) 574-9888
e-mail: mansfieldbk@ornl.gov • website: www.ornl.gov/hgmis

The U.S. Human Genome Project is the fifteen-year national coordinated effort to discover and characterize all of the estimated 80,000–100,000 genes in human DNA and render them accessible for further biological study. The program will also address the ethical, legal, and social issues that may arise from the project. It publishes the newsletter *Human Genome News* and several documents, which include *Your Genes, Your Choices* and *Department of Energy Primer on Molecular Genetics*.

Institute for Global Communication (IGC)
18 De Boom St., San Francisco, CA 94107
(415) 442-0220 • fax: (415) 546-1794
e-mail: support@igc.apc.org • website: www.igc.org

The institute provides computer networking services for international communications dedicated to environmental preservation, peace, and human rights. IGC networks include EcoNet, ConflictNet, LaborNet, and PeaceNet. It publishes the monthly newsletter *NetNews*.

Interactive Services Association (ISA)
8403 Colesville Rd., Suite 865, Silver Springs, MD 20910
(301) 495-4955
e-mail: isa@aol.com • website: www.isa.org

ISA is a trade association representing more than three hundred companies in advertising, broadcasting, and other areas involving the delivery of telecommunications-based services. It has six councils, including Interactive Marketing and Interactive Television, covering the interactive media industry. Among the association's publications are the brochure *Child Safety on the Information Superhighway*, the handbook *Gateway 2000*, the monthly newsletter *ISA Update*, the biweekly *Public Policy Update*, and the *ISA Weekly Update* (delivered by fax or e-mail).

International Computer Security Organization (ICSA)
1200 Walnut Bottom Rd., Carlisle, PA 17013
(717) 258-1816 • (800) 488-4595 • fax: (717) 243-8642
website: www.icsa.net

ICSA offers information and opinions on computer security issues. It strives to improve computer security by disseminating information and certifying security products. The association publishes the bi-monthly *ICSA Newsletter*.

International Society for Technology in Education (ISTE)
University of Oregon, 1787 Agate St., Eugene, OR 97403
(800) 336-5191 • fax: (503) 346-5890
e-mail: iste@iste.org • website: www.iste.org

ISTE is a multinational organization composed of teachers, administrators, and computer and curriculum coordinators. It facilitates the exchange of information and resources between international policy makers and professional organizations related to the fields of education and technology. The society also encourages research on and evaluation of the use of technology in education. It publishes the journal *Computing Teacher* eight times a year, the newsletter *Update* seven times a year, and the quarterly *Journal of Research on Computing in Education*.

National Library of Education
555 New Jersey Ave. NW, Room 101, Washington, DC 20208-5721
(800) 424-1616 • fax: (800) 219-1696
e-mail: Library@inet.ed.gov • website: www.ed.gov

The library provides specialized subject searches and retrieval of electronic databases. Its other services include document delivery by mail and fax, research counseling, bibliographic instruction, interlibrary loan services, and selective information dissemination. For those who have access to the Internet, the library provides general information about the Department of Education, full-text publications for teachers, parents, and researchers, and information about initiatives such as *GOALS 2000, Technology*, and *School-to-Work Programs*.

National School Boards Association (NSBA)
1680 Duke St., Alexandria, VA 22314
(703) 838-6722 • fax: (703) 683-7590
e-mail: info@nsba.org • website: www.nsba.org/itte

The association is a federation of state school boards. NSBA advocates equal opportunity for primary and secondary public school children through legal counsel, research studies, programs and services for members, and annual conferences. It also provides information on topics such as curriculum development and legislation that affects education. NSBA endorses the use of computers as an educational tool. The association publishes the bimonthly newsletter *A Word On*, the monthly *American School Board Journal*, the biweekly newspaper *School Board News*, and numerous other publications.

Special Interest Group for Computers and Society (SIGCAS)
c/o Association for Computing Machinery
1515 Broadway, 17th Floor, New York, NY 10036
(212) 869-7440 • fax: (212) 944-1318
e-mail: infodir_sigcas@acm.org • website: www.acm.org/sigcas

SIGCAS is composed of computer and physical scientists, professionals, and other individuals interested in issues concerning the effects of computers on society. It aims to inform the public of issues concerning computers and society through such publications as the quarterly newsletter *Computers and Society*.

Bibliography of Books

Alan B. Albarran and David H. Goff — *Understanding the Web: The Social, Political, and Economic Dimensions of the Internet.* Ames, IA: Iowa State University Press, 2000.

Cynthia J. Alexander and Leslie A. Paul — *Digital Democracy: Policy and Politics in the Wired World.* New York: Oxford University Press, 1998.

Martin Bichler — *The Future of E-Markets: Multi-Dimensional Market Mechanisms.* Cambridge, MA: Cambridge University Press, 2001.

David B. Bolt and Ray Crawford — *Digital Divide: Computers and Our Children's Future.* New York: TV Books, 2000.

Clay Calvert — *Voyeur Nation: Media, Privacy, and Peering in Modern Culture.* Boulder, CO: Westview, 2000.

Jeremy R. Chapman, Celia Wright, and Mark Deierhoi — *Organ and Tissue Donation for Transplantation.* New York: Oxford University Press, 1997.

Michael Chesbro — *The Complete Guide to E-Security: Using the Internet and E-Mail Without Losing Your Privacy.* Boulder, CO: Paladin, 2000.

Lakshmanan Chidambaram and Ilze Igurs — *Our Virtual World: The Transformation of Work, Play, and Life Via Technology.* Hershey, PA: Idea Group, 2001.

April Christofferson — *Clinical Trial.* New York: St. Martin's, 2000.

Michael A. Civin — *Male, Female, Email: The Struggle for Relatedness in a Paranoid Society.* New York: The Other, 2000.

Benjamin Compaine — *The Digital Divide: Facing a Crisis or Creating a Myth.* Cambridge, MA: MIT Press, 2001.

David K. Cooper and Robert P. Lanza — *Xeno: The Promise of Transplanting Animal Organs into Humans.* New York: Oxford University Press, 2000.

Edward Cornish — *Exploring Your Future: Living, Learning and Working in the Information Age.* Boston, MA: World Future Society, 1996.

Simson Garfinkel — *Database Nation: The Death of Privacy in the 21st Century.* Cambridge, MA: O'Reilly & Associates, 2001.

Brandon Garrett — *The Right to Privacy.* Brookshire, TX: Rosen, 2000.

Kenneth Getz — *Word from Study Volunteers: Opinions and Experiences of Clinical Trial Participants.* Boston, MA: CenterWatch, 2001.

Charles Grantham	*The Future of Work: The Promise of the New Digital Work Society*. New York: McGraw-Hill, 1999.
Edward E. Hindson and Lee Frederickson	*Future Wave: End Times, Prophecy, and the Technological Explosion*. Eugene, OR: Harvest House, 2001.
Michael S. Hyatt	*Invasion of Privacy*. Washington, D.C.: Regnery, 2001.
Merritt Ierley	*The Comforts of Home: The American House and the Evolution of Modern Convenience*. British Columbia: Crown, 2001.
Dianne N. Irving	"NIH and the Human Embryo Revisited: What's Wrong with This Picture?" *American Bioethics Advisory Commission*, 1999.
Bart Kosko	*The Fuzzy Future: From Society and Science to Heaven in a Chip*. British Columbia: Crown, 1999.
Straubhaar Larose	*Electronic Media in the Information Age*. Stamford, CT: Thomson Learning, 2001.
Michael Lewis	*Next: The Invisible Revolution*. New York: Random House, 2001.
Darcy Lockman	*Kaleidoscope—Technology: Computers, Robots, the Internet, Computer Animation*. New York: Marshall Cavendish, 2001.
Scarlet McGwire	*Surveillance: The Impact on Our Lives*. Orlando, FL: Raintree Steck-Vaughn, 2001.
Elizabeth Neill	*Rites of Privacy and the Privacy Trade: On the Limits of Protection for the Self*. Canada: McGill-Queens University, 2001.
Alan L. Porter and William H. Read	*The Information Revolution: Current and Future Consequences*. Westport, CT: Greenwood, 1998.
Jeffrey Rosen	*The Unwanted Gaze: The Destruction of Privacy in America*. New York: Random House, 2001.
David M. Safon	*Workplace Privacy: Real Answers and Practical Solutions*. Stamford, CT: Thomson Learning, 2000.
Andrew L. Shapiro	*The Control Revolution: How the Internet Is Putting Individuals in Charge and Changing the World We Know*. New York: Perseus, 1999.
James Slevin	*The Internet and Society*. United Kingdom: Polity, 2000.
Mark L. Smith	*Managing the Internet Controversy*. New York: Neal-Schuman, 2000.

Bethany Spielman   *Organ and Tissue Donation: Ethical, Legal, and Policy Issues*. Carbondale, IL: Southern Illinois University Press, 1996.

Mark J. Stefik   *The Internet Edge: Social, Technical, and Legal Challenges for a Networked World*. Cambridge, MA: MIT, 1999.

Craig Warkentin   *Reshaping World Policies*. Boston: Rowman & Littlefield, 2001.

William Webb   *The Future of Wireless Communications*. Norwood, MA: Artech House, 2001.

Index